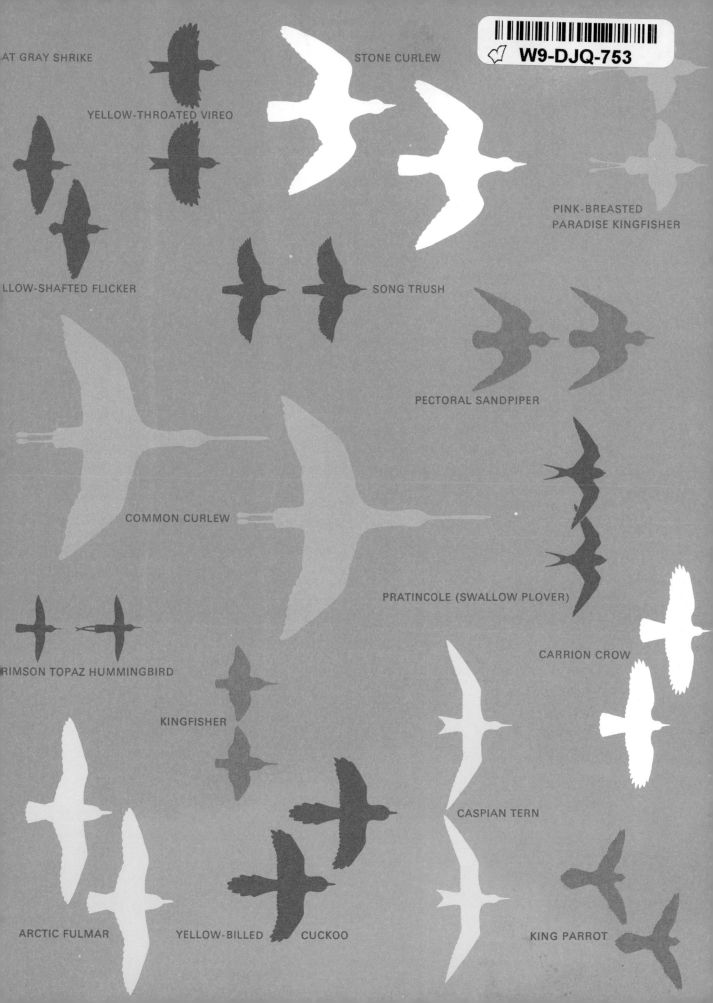

AT GRAY SHRIKE

YELLOW-THROATED VIREO

STONE CURLEW

PINK-BREASTED
PARADISE KINGFISHER

LLOW-SHAFTED FLICKER

SONG TRUSH

PECTORAL SANDPIPER

COMMON CURLEW

PRATINCOLE (SWALLOW PLOVER)

CARRION CROW

RIMSON TOPAZ HUMMINGBIRD

KINGFISHER

CASPIAN TERN

ARCTIC FULMAR

YELLOW-BILLED   CUCKOO

KING PARROT

MAGNIFICENT FRIGATE-BIRD

BEAUTIFUL NUTHATCH

GOSHAWK

OYSTERCATCHER

RUFOUS HUMMINGBIRD

IVORY-BILLED
WOODPECKER

EURASIAN SPOONBILL

GRAY FANTAIL

MAGPIE

COMMON SWIFT

RING-NECKED PHEASANT

SHELDUCK

HERRING GULL

HAWAIIAN PETREL   WHITE STORK   MUSSCHENBROEK'S LORIKEET   KNOT

THE FAMILY LIFE OF BIRDS

BLACKBURNIAN WARBLER

RUDDY TURNSTONE

PRAIRIE CHICKEN

COMMON SNIPE

BEE-EATER

ROYAL FLYCATCHER

GANNET

WHITE-CHINNED PETREL

GOLDEN-FRONTED HANGING PARROT

PENNANT-WINGED NIGHTJAR

BLACKBIRD   CHIFFCHAFF   BLACKCAP   CIRL BUNTING   GREAT NIGHTINGA[

LESSER GRAY SHRIKE   AMERICAN ROBIN   FIELDFARE

ICTERINE WARBLER   BONELL

PHOTOGRAPHS AND TEXT BY
HANS D. DOSSENBACH

CONCEIVED AND DESIGNED BY
EMIL M. BÜHRER

THE FAMILY LIFE OF BIRDS

FOREWORD AND SCIENTIFIC ADVISOR
PROF. OTTO KOENIG

McGRAW-HILL BOOK COMPANY
NEW YORK ST. LOUIS SAN FRANCISCO TORONTO

GREAT REED WARBLER   REED WARBLER   MISTLE THRUSH   BLACK-THROATED JAY   MAGPIE
OCK   STARLING   YELLOWHAMMER   LESSER GRAY SHRIKE
LER

THE FAMILY LIFE OF BIRDS

A McGRAW-HILL CO-PRODUCTION
TRANSLATED FROM THE GERMAN
BY FRITZ BAUCHWITZ
AMERICAN CONSULTANT:
BERTEL BRUUN
SCIENTIFIC ADVISOR: OTTO KOENIG
EDITORIAL: NANCY KELLY
PRODUCTION ASSISTANT:
FRANCINE PEETERS
DIAGRAMS AND CHARTS
BY FRANZ CORAY

ALL OTHER DRAWINGS
BY ROLF BAUMANN AND RUDOLF KUENZI
PRINTED AND BOUND
BY C. J. BUCHER AG, LUCERNE
SWITZERLAND

PIGMY FALCON   HOUSE SPARROW   SKY LARK   LESSER SHORT-TOED LARK

SONG THRUSH   SUPERB GLOSSY STARLING   GOLDEN ORIOLE

OLD WORLD KESTREL

ABOUT THIS BOOK
By Dr. Otto Koenig

Somewhere there must be a world without insecticides, without air pollution and without excessive radioactivity, a world where the sky is still as blue as it was a thousand years ago, without haze and without smog. I do not know where that world might be—but Hans Dossenbach evidently has discovered it.

This seems to be borne out by his pictures. For him, and for his readers, the world is still intact. One is almost tempted to regard him as a last survivor of the Stone Age—one who set out to create a new Altamira with his camera, and to celebrate a worldwide quarry, which is unaware that it has been overrun by civilization and continues to populate its various habitats as gaily, as colorfully, as nervously, and as calmly as it lives in the stirring pictures of this book.

I do not know anyone who photographs as Dossenbach does. His lenses obviously function differently from other optical systems: the camera, as well as being an instrument for depicting what he sees, serves him as an interpretative vehicle for his thoughts and dreams. There is no other explanation for the strangely lyrical quality of his pictures. He turns every photograph into a story, an account of nature.

I first met Dossenbach at Lake Neusiedl. Wearing a faded bush shirt, he went about his work as if nothing existed for him but the one task at hand and the particular bird's nest he was about to photograph. Nearly all photographers will eliminate any stalks that screen the nests in order to get an unobstructed view. Dossenbach carefully bent them aside. When his work was done, everything looked as if no one had ever been there to discover or observe anything. And so each of his pictures shows that none of the animals could possibly have been aware of the photographer's existence.

No greater compliment can be paid to an animal photographer. In our completely sick world, Dossenbach's pictures help us to regain our belief in the wisdom of nature because they ignore man, who has brought about that sickness and continues to spread it. Anyone who leafs through this book will soon find his pleasure mounting from page to page.

GREAT GRAY SHRIKE   RED-FOOTED FALCON   BULLFINCH

WOODCHAT SHRIKE   ROBIN   HOUSE SPARROW

GUILLEMOT   CHINESE THRUSH

GYRFALCON

SPOTTED FLYCATCHER

LESSER WHITETHROAT

RED-BACKED SHRIKE

LESSER KESTREL

WOOD THRUSH

EMU

BLACK GUILLEMOT

CRESTED LA

Directly in front of me, in a cage on the desk, while sitting over one of the texts for the present book, two tiny, colorful birds, whose original habitat is Australia, were suddenly overcome by the mating urge. They demonstrated a tender affection for one another, began to court each other fervently, and impetuously consummated their union again and again. They then collected coir fibers and cotton threads scattered on the floor of the cage and built a nest with them. And, barely two weeks after their mating, the first egg, white and just a little larger than a pea, was lying in the pretty, softly padded bowl of the nest. Another one followed every day until there were five eggs in all. The male and female alternated in sitting on their eggs, an arm's length away from the pounding typewriter, passing on the feverish heat of their little bodies (which weigh just over two tenths of an ounce each), and bringing the eggs to life. Bursting with vitality at other times, those little birds displayed an almost incomprehensible patience during the twelve days of incubation. Only when the pink, naked, blind little nestlings had broken through the shells and demanded food with wide-open bills and barely audible peeping sounds, did the parents resume their restless activity. And two weeks later, the brood was already flying about in the cage, differing from their parents only in their plain adolescent plumage. They had grown to full size, looked for their own food, bathed exuberantly in the water bowl, preened their feathers painstakingly— in short, they had developed into "responsible" birds and were, for all intents and purposes, independent.

Anyone who has experienced such a development just once is enthralled by it. And the more intense his involvement, the greater will be his amazement, his desire to learn more about it, and to seek the solution of the mysteries surrounding birds.

The ornithologist—as the bird researcher is called—has dissected their light bodies and investigated them to the last fiber. Geneticists, endocrinologists, and behavioral scientists have tried to unriddle the mystery of reproduction, the most important event in the life of the birds. But we can come to understand a great deal about the mating life of these animals simply by going out on a Sunday morning and observing the tricks of our feathered friends through field glasses—or, better yet, by letting them affect us with their wanderlust and following them to their breeding grounds. Here we can observe something quite overwhelming: the life cycle of these marvelous creatures.

There is their incredible adaptability, which enables them to build their nests wherever nature offers the prerequisites for sustaining life, up to the extreme limitations of possible survival. There is their supreme mastery of nest construction and the architectural boldness of the structures themselves. There is their tremendous social mobility at mating time. There is their mating mood, accompanied by the most peculiar and delightful courtship rituals. There is the captivating enchantment of their mating plumage. There is the great variety of their breeding habits and liaisons: we find among them confirmed monogamists, passionate polygamists, and unbridled polyandrists; there are unions that last for only a few moments, some that last for a single breeding season, and still others that last for life. And within these broad mating patterns, each species has its own peculiar rituals and customs.

We encounter their amazing eggs, concealing the greatest of all miracles: the development of life. We observe the breeding habits and the brooding behavior of the bird parents which so often strike us as strange. We follow the development of the new life within the protective envelope of the egg and the laborious squirming of the young chick as it emerges from the egg without assistance from its parents and everything that occurs from then on in connection with the newly hatched offspring. And we can witness the problems involved in raising and maintaining a brood, problems which most of us probably did not even know to exist.

On the following pages: Flamingos in the Sechura desert in Northern Peru. These birds live and breed along shallow salt lakes located in the murderous heat of the rift valley or in the extremely harsh climate of the Andes plateau.

REED BUNTING

CHOUGH

GREAT TIT

OSPREY

NIGHTINGALE

WHINCHAT

TREE SPARROW

SPOTTED NORTHURA

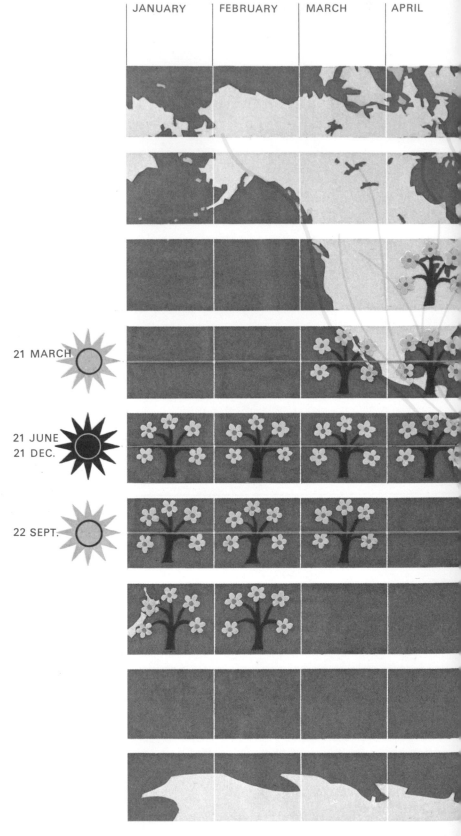

The life rhythm of birds is dependent on the seasons. For most species the prerequisites for thriving offspring are summer heat and an ample food supply containing plenty of proteins.

Most birds who live in the equatorial area are capable of brooding all year round and when they migrate at all they travel only short distances to evade heavy rainfall. The farther north or south of the equator they live, however, the more precisely they must adjust their incubation periods to the approach of warm weather. As is the case with North American, European, and North Asiatic species, there are migratory species among the birds that breed in South Africa and South America. These avoid the cold season by moving to warmer regions. And, like some of the storm-inured breeders of the Arctic who move 9,400 miles to the Antarctic when autumn comes, the Antarctic birds migrate to the far north in April to spend their resting period there.

But some species pay no heed to the seasons. Emperor penguins lay their eggs during the southern autumn and hatch and coddle their young during the murderous, dark winter months. Some species of owls will occasionally sound their mating calls in December and sit on their eggs in early February, while crossbills frequently build their nests in the snow-covered woods of January.

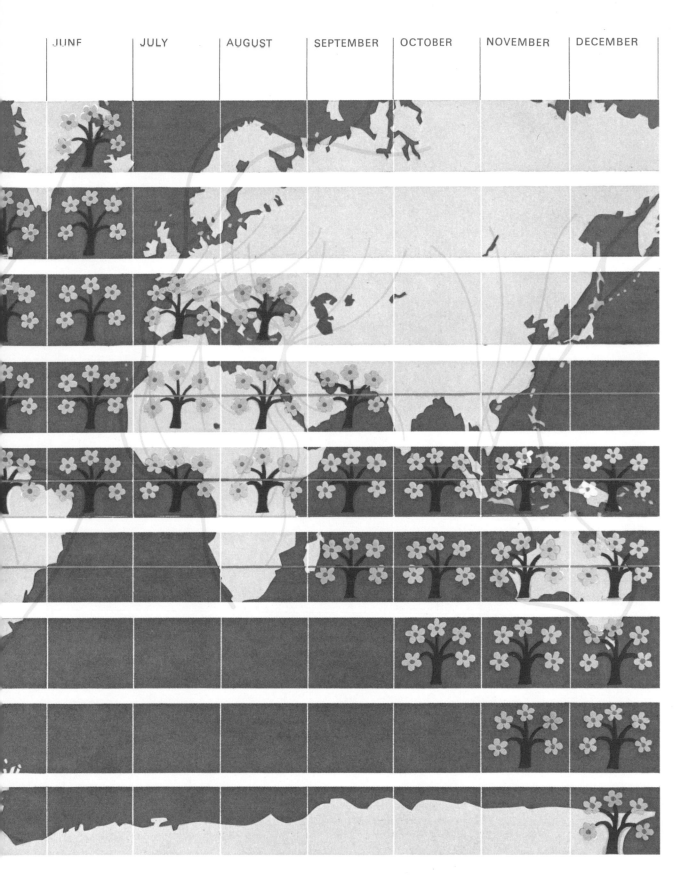

The blossoming trees on the map signify the major breeding seasons in the various regions of the world. The variously colored lines indicate important flight routes of migratory birds. They show that many of the migrants—especially waders, gulls and terns—keep to the coastlines, while the majority of land birds migrate in a broad front and, as a rule, attempt to travel over land areas wherever possible (over isthmuses the formations can be observed to close up as if they were passing through an invisible funnel).

THE ROCKY BIRD ISLE IN THE NORTH

Of all the migratory species, the sea birds are the most restless. Some of them loners, others in myriad flights, they crisscross above the oceans in search of fish. Their unbridled wanderlust seems to be unaffected by even the slightest trace of nostalgia. But in the lengthening days and shortening nights of spring they all are suddenly drawn to their breeding grounds. They are seized by the stirring of the mating instinct which leads them to surf-battered rocks, to lonely islets, and sandy beaches.

Since the previous summer these spots have been deserted and desolate, visited only by a few feathered strays and the storms of autumn and winter. Even the plateaux, ledges, and crevices, white with calciferous excrement, scarcely suggest the multitudes that congregate here at mating time.

At first there are a few early arrivals, then they swarm in by the thousands: auks, murres, cormorants, terns, gulls, boobies, and tubenoses in an unbelievable confusion. It scarcely seems possible that eventually they will sort themselves out in a precise, orderly arrangement. Indeed, even after this order has been estab-

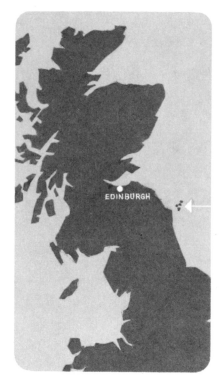

The Farne Islands near the east coast of England.

Right: The cramped space on the remote rocks of the Farne Islands provides breeding quarters for countless birds. The plateau is covered with common murres while kittiwakes brood in the walls.

On the following pages: Every inch of nesting space on the islands is utilized. The larger projections on the cliffs are occupied by shags with their spacious seaweed nests. The kittiwakes are satisfied with the tiniest bays for the construction of their remarkably solid grass nests, which they plaster with excrement and cement to the rock during the breeding period.

lished, the outsider would hardly recognize it as such. And yet it is already in force within the very first days: each male bird immediately begins to court the first female of his species he encounters; he postures before his bride-elect, intimidating with unmistakable threatening gestures any and all who dare to encroach upon his domain. The great cormorants, each standing the ground he has conquered, beckon enticingly with their wings and give signals with the spots that have turned white at mating in the plumage of their thighs. Auks and murres bow devotedly and exhibit swaying, waddling strides. Gulls and terns punctuate their gestures with strident screeches. Once chosen, the breeding site, which frequently is no more than a wingspan away from the next one, is defended against all intruders and joins the couple in an inseparable bond for the next few months. Each occupant unfailingly can identify, among the thousands of nesting sites, the tiny area where he will propagate his kind. The eggs, which all appear to be identical to human eyes, are readily distinguishable to—and fiercely protected by the nesting parents.

Mating is not an easy proposition for the shag. From the very start he is faced with problems which will tax his ingenuity considerably. Abandoning his hitherto unfettered, piratical existence, the male shag must settle down to a designated task, a task which will absorb him completely, day and night.

First he must find a suitable nesting site. As is customary with

He must lose no time, because very soon even the narrowest projection and smallest niche will have a tenant. In this precarious position he waits, head thrown back, quacking, and wings flapping in an animated effort to entice a passing lady shag.

It is soon evident that the male shag is as choosy as the female, perhaps even more so. He is rarely satisfied with the first to come

acceptable to her lord and master she must demonstrate her humility by means of very definite and complicated body movements, movements she has already begun during her approach. If the male is seriously interested, he will initiate a rather raucous welcoming ceremony in which the female participates vigorously. The latter cannot consider herself to be engaged, however, until she is left alone at the nesting site while

northern cormorants—and the shag is a member of this ilk—he will select a narrow rock ledge and stand with his broad waddling feet on a small projection on which he can barely turn around.

along. Again and again, a female, captivated by his noisy lovesong, will land near him only to be promptly chased away. Frequently, she is even brutally pushed off the projection. To make herself

the male goes off in search of construction materials. Conscious of her responsibility, she will doggedly defend the site against all strangers in his absence. When he finally returns, triumphantly bran-

Only when the male shag presents his partner with the first bunch of seaweed after the initial courting gestures, may she consider herself his bride, for he is choosy and almost never satisfied with the first comer. Once the couple has reached an agreement, she will pass detailed judgment on the nesting site, which he has selected and which usually consists of a small rocky projection. A few days after their first encounter, the couple will begin nest construction; he will fetch almost all the necessary materials, seaweed and grass, which she will weave, preferably while wet, into a clumsy but remarkably solid nest. The nesting site is obstinately defended against all intruders, and even the arduously assembled construction materials must be constantly guarded by one of the partners.

dishing a bunch of seaweed in his beak, she immediately starts building the nest.

Other cormorants, which brood on flat ground, build rather untidy nests as a rule, and some may even be satisfied with a pile of their excrement, but the home of our shags will be handsome and solid. Indefatigably the male searches out the needed materials. In addition to the seaweed he also brings in tufts of grass from the mainland, after first dipping them in seawater. His mate devotes herself entirely to the construction of the cradle for the offspring. Adroitly she intertwines the nesting materials with her beak, occasionally using her feet

to hold on the excess. In due course the structure is cemented to the cliff with the abundant excrement to be found on the rocks. Shags, like all cormorants, have ample time to devote to nest construction during the day. They are such excellent underwater fishers that, under favorable conditions, they are able to catch their daily rations within half an hour, a feat that can be equalled by hardly any other bird. Construction progresses swiftly, and in a short time the nest is sturdy enough to accommodate the eggs. Our shags, however, obviously enjoy their architectural task, because they will continue tinkering with their home as long as they live in it.

Both parents participate with equal devotion in the care of their offspring, spelling each other during the four weeks of the incubation process and the five weeks during which the chicks require care. Each brooding relief follows

a designated ritual, a succession of peculiar motions whose significance is not completely known to ornithologists. The departure of the tending partner from the nest also is accompanied by a

carefully prescribed ceremony. The bird stands with its back to the sea, inclining its extended neck diagonally to the rear. After a few additional ritual motions it spreads its wings and soars away. If the departing mate were to fly up abruptly without "telling" his brooding partner—as would be the case in event of sudden danger—the other might be frightened into headlong flight instead of coming to relieve its mate. This

would expose the eggs or the nestlings to those constantly patrolling predatory birds, the greater or lesser black-backed gulls. Even for short periods such exposure could be disastrous.

Every vacancy on the bird island is occupied, all opportunities are exploited.

Full of distrust, roseate terns and eider ducks hide their isolated nests in the sparse, low vegetation which manages to thrive on some of the islands. Arctic terns and Sandwich terns, on the other hand, can afford to brood entirely in the open, since they form populous colonies in which a sufficient number of birds is always prepared to rush, screeching, at the rapacious larger gulls and drive them off with savage aerial attacks. Any tolerably flat area may serve as a nesting site, be it a sand or gravel bar, a strip of short grass or a level rock.

Auks and murres seek greater protection. They prefer to lay their eggs on the narrow stone ledges of sheer, all but inaccessi-

ble cliffs. Where they live on plateaus, these are always protected by steep precipices. Cormorants and fulmars nest in recesses and on projections of rocks, and kittiwakes expertly attach their grass nests on the narrowest of rock balconies, using excrement as reinforcing plaster.

The egg of the razor-billed auk could hardly roll off the narrow rock ledge, even if the sitting bird should take headlong flight. Once in motion, its extreme conical shape would cause it to roll in a small circle. Below: In contrast to most of their relatives, kittiwakes brood on rocky projections.

The Great Breeding Areas of the World

SCANDINAVIA provides many swamp and game areas favored by shorebirds, geese, and ducks, birds of prey, and owls, as well as quiet coastal areas and islands populated by many thousands of sea birds.

It is obviously impossible to describe here all the world's significant breeding grounds. We must, therefore, limit ourselves in this brief text to a cursory survey of those areas on each continent which are of the greatest ornithological importance. The map of the following pages, however, points out all the world's major breeding sites.

AUSTRIA's large (but unfortunately seriously threatened) reed forests of the Neusiedler Lake are famous for their extensive colonies of great white herons, purple herons, gray herons, and spoonbills. Numerous rails, wading birds, ducks, and graylag geese, as well as white storks and small reed dwellers also breed in this vicinity, and the great bustard is found in the adjoining dry land areas. The rare black stork breeds along the Danube and March Rivers.

◀ *A foolish guillemot is looking after two children of its neighbors along with its own young. The peculiar spectacle markings are found only on certain members of the guillemot family. The photograph was taken on the Farne Islands, one of the numerous sea bird breeding grounds of Great Britain.*

GREAT BRITAIN is notable for its numerous sea bird breeding areas, most of them rocky islands, some of which are visited by incredible numbers of birds during the breeding season. Nearly 8,000 eider ducks and thousands of other birds of various species breed on the Farne Islands along the east coast of England. Bass Rock in Scotland is invaded yearly by some 20,000 gannets, while nearly 100,000 gannets and about 50,000 fulmars breed on the St. Kilda Islands. In addition, many wading birds nest in the marshy areas and among the dunes along Great Britain's extensive shoreline.

HOLLAND and BELGIUM provide some of the most important breeding areas in Europe located in lagoons and marshes and on several islands, for such aquatic and sea birds as geese, ducks, and terns, gray herons, cormorants, and spoonbills. Aquatic birds by the hundreds of thousands assemble here for the winter.

SPAIN and PORTUGAL, in the extensive marshy deltas of the Guadalquivir, Ebro, and Tejo Rivers, provide noteworthy breeding sites for various herons, white storks, numerous aquatic birds, and such rare birds of prey as the imperial eagle and the red kite.

ICELAND provides breeding sites for many thousands of sea birds along its coast and for a huge number of ducks and geese around the lakes of the interior.

SWITZERLAND's National Park, in the southeast part of that landlocked nation, attracts an interesting population of brooding mountain birds, among them capercaillies, black grouse, rock partridge, hazel hens, ptarmigan, golden eagles, nutcrackers, and ravens. There are also several smaller reed areas that shelter marsh dwellers.

FRANCE has a significant breeding area in the Camargue, the partly brackish alluvial land of the Rhône delta where more than 10,000 herons of several species, thousands of flamingos, and numerous wading birds gather each season. It is also a wintering area for more than 150,000 ducks. Several other important breeding areas for wading and sea birds are located in the north and northeast parts of the country.

EASTERN EUROPE includes some countries with important, and for the most part carefully tended, breeding areas. The interior of the Soviet Union, of Poland, Czechoslovakia and Hungary provide many reed-grown and swampy lakes for such wading bird as spoonbills, a variety of herons, ibises, and black and white storks. Among the outstanding species of these regions are the red-breasted goose, the lesser spotted eagle, the osprey, the imperial eagle, and the white-tailed sea eagle. Along with important wintering areas for geese, ducks and shorebirds, the Balkans contain the only European breeding grounds of the eastern white pelican and the Dal-

GERMANY has a number of sea bird sanctuaries, primarily for terns and gulls, along the North and the Baltic sea coasts, while various lake areas in the interior provide breeding sites for reed dwellers. Lake of Constance is a favorite breeding area for red-crested pochards and black-necked grebes, and the Lausitz Ponds in East Germany are particularly hospitable to black storks, rollers, and cranes. Golden eagles, capercaillies, black grouse, hazel hens, and rock partridge, among others, breed in the mountainous countryside around the Königssee in Bavaria.

matian pelican. The largest of these grounds are located at the Danube delta in Rumania.

NORTH AMERICA. The Everglades in Southern Florida, with their huge saw grass swamps and mangrove forests, provide the United States with one of the most remarkable breeding areas in the Western Hemisphere. The many species of birds found there include the roseate spoonbill, the white ibis, the wood stork, the limpkins, and the Everglades kite, along with black skimmers, and brown pelicans. In addition, there are innumerable transients and winter residents.

Important breeding areas—swamps, lakes, and evergreen forests—are located in Southern Oregon, which is inhabited primarily by ducks, geese, American white pelicans, double-crested cormorants, and various herons. Several millions of transient ducks nest in this area.
Canadian geese, wood ducks, great northern divers, sandhill cranes, bald eagles, ruffed grouse, and numerous other species brood in the Manistee Swamps of Michigan, primarily near artificial lakes and ponds.

Canada has particularly vast breeding areas. One of the world's most populous duck breeding grounds is situated in Prince Albert National Park, which extends over nearly 1,540 square miles. 250,000 snow geese and 200,000 fulmars, among other species, brood on Baffin Island. The few surviving American whooping cranes—among the rarest birds of all—breed in Wood Buffalo National Park in Central Canada.
Other colonies—some small and some extremely large—of brooding sea birds can be found in many of North America's areas, particularly in Canada.
Several sanctuaries in Hawaii provide breeding grounds for Pacific sea birds, for instance about 300,000 Laysan albatrosses and black-footed albatrosses, various boobies and countless terns, along with the extremely rare Hawaiian goose, and some rare land-based birds indigenous to the area.

AFRICA. The bird life of eastern and southeastern African waters is among the world's most abundant. About a hundred species are to be found along the rivers and more than five hundred species have been identified in certain areas. The saddle bill stork, the African wood ibis and the open bill stork, along with the Goliath heron, the sacred ibis, the African sea eagle, African darter and various cormorants, and Egyptian geese, tree ducks, rails, and kingfishers are among the most conspicuous.

Most of these species also are found near the numerous shallow salt lakes, but along some of these waters of the Rift Valley, the scene is

IMPORTANT BREEDING AREAS IN NORTH AMERICA

CANADA

1 BOWMAN BAY (BUFFIN ISLAND)
2 ESKIMO POINT, KEEWATIN (NORTHWEST TERRITORIES)
3 CAPE SEARLE (BUFFIN ISLAND)
4 PERRY RIVER (NORTHWEST TERRITORIES)
5 WOOD BUFFALO NATIONAL PARK (ALBERTA)
6 CAP TOURMENTE (NEW BRUNSWICK)
7 PRINCE ALBERT NATIONAL PARK (SASKATCHEWAN)
8 BONAVENTURE ISLAND (QUEBEC)

UNITED STATES

9 ST. LAURENT ISLAND (ALASKA)
10 KUSKOKWIM WILDLIFE RANGE (ALASKA)
11 IZEMBEK WILDLIFE RANGE (ALASKA)
12 MOUNT MCKINLEY NATIONAL PARK (ALASKA)
13 NATIONAL WILDLIFE REFUGES, SOURIS RIVER (NORTH DAKOTA)
14 NATIONAL WILDLIFE REFUGES IN KLAMATH BASIN (OREGON, CALIFORNIA)
15 SENEY NATIONAL WILDLIFE REFUGE (MICHIGAN)

16 MONOMOY NATIONAL WILDLIFE REFUGE (MASSACHUSETTS)
17 PARKER RIVER NATIONAL WILDLIFE REFUGE (MASSACHUSETTS)
18 BRIGANTINE NATIONAL WILDLIFE REFUGE (NEW JERSEY)
19 BEAR RIVER MIGRATORY BIRD REFUGE (UTAH)
20 CRAB ORCHARD NATIONAL WILDLIFE REFUGE (ILLINOIS)
21 CHINCOTEAGUE NATIONAL WILDLIFE REFUGE (VIRGINIA, MARYLAND)
22 GREAT SMOKY MOUNTAINS NATIONAL PARK (TENNESSEE, NORTH CAROLINA)
23 SALT PLAINS NATIONAL WILDLIFE REFUGE (OKLAHOMA)
24 CAPE ROMAIN NATIONAL WILDLIFE REFUGE (SOUTH CAROLINA)
25 IMPERIAL NATIONAL WILDLIFE REFUGE (CALIFORNIA)
26 SAGUARO NATIONAL MONUMENT (ARIZONA)
27 LAGUNA ATASCOSA NATIONAL WILDLIFE REFUGE (TEXAS)
28 SANTA ANNA NATIONAL WILDLIFE REFUGE (TEXAS)
29 LOXAHATCHEE NATIONAL WILDLIFE REFUGE (FLORIDA)
30 HALEAKALA NATIONAL PARK (HAWAII)
31 HAWAIIAN ISLANDS NATIONAL WILDLIFE REFUGE
32 WILLAPA NATIONAL WILDLIFE REFUGE (WASHINGTON)
33 GLACIER NATIONAL PARK (MONTANA)
34 YELLOWSTONE NATIONAL PARK (WYOMING, MONTANA, IDAHO)
35 TOBY POND (LONG ISLAND)
36 JAMAICA BAY WILDLIFE REFUGE (NEW YORK)

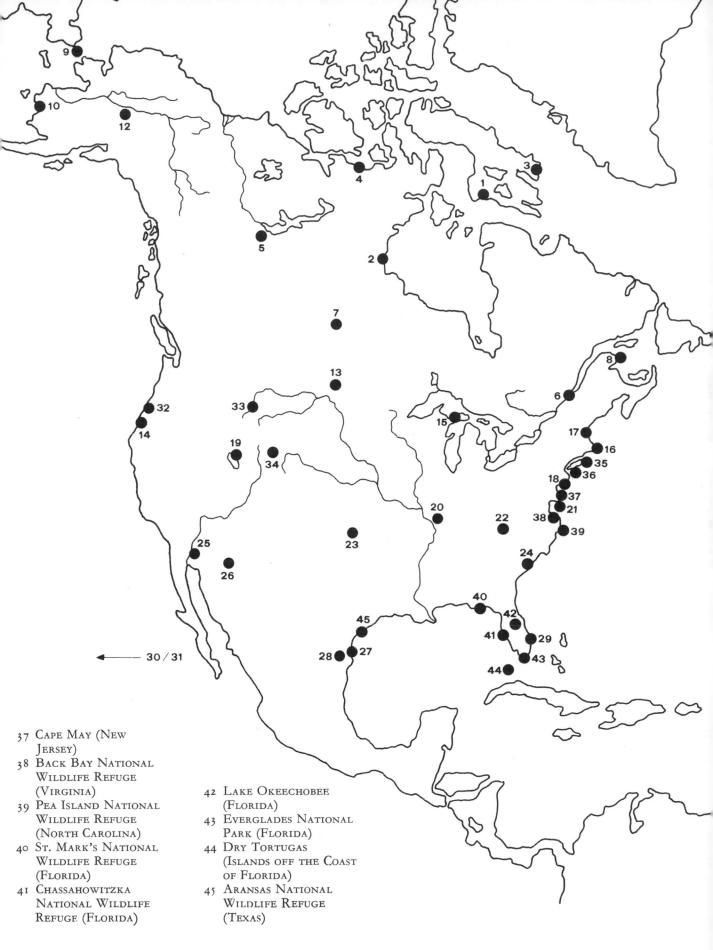

37 CAPE MAY (NEW
  JERSEY)
38 BACK BAY NATIONAL
  WILDLIFE REFUGE
  (VIRGINIA)
39 PEA ISLAND NATIONAL
  WILDLIFE REFUGE
  (NORTH CAROLINA)
40 ST. MARK'S NATIONAL
  WILDLIFE REFUGE
  (FLORIDA)
41 CHASSAHOWITZKA
  NATIONAL WILDLIFE
  REFUGE (FLORIDA)

42 LAKE OKEECHOBEE
  (FLORIDA)
43 EVERGLADES NATIONAL
  PARK (FLORIDA)
44 DRY TORTUGAS
  (ISLANDS OFF THE COAST
  OF FLORIDA)
45 ARANSAS NATIONAL
  WILDLIFE REFUGE
  (TEXAS)

dominated by the flamingos. As many as three million greater and lesser flamingos gather at times at Lakes Nakuru and Magadi in Kenya. On occasion, these bodies of water seem to be covered with a huge rose-colored carpet.

In the steppes and the bush, too, one constantly comes upon such conspicuous examples of bird life, as the ostrich, the crowned crane, the secretary bird, the hornbill, the weaverbirds and wax bills, bustards, guinea fowl and francolines, in addition to such jewel-like creatures as the lilac-breasted rollers and various bee-eaters.

Along the Mediterranean coast of North Africa there are a number of noteworthy heron colonies. Myriad herons, pelicans, flamingos, cormorants, gulls, and terns breed on Beau d'Arguin, an island in Mauritania, which was rediscovered only a few years ago. Huge sea bird colonies with many hundreds of thousands of cape cormorants and jackass penguins are also found on the islands along the coast of South Africa, which maintains the world's largest artificial breeding colony for wild birds—a wooden platform with an area of 185,000 square feet, where each year 100,000 cormorants leave 990 tons of excrement—a valuable plant fertilizer—during the breeding season.

SOUTH AMERICA is called "the continent of birds" for good reason. A greater variety of species breeds nowhere else in the world. To date, more than 1,550 different species have been identified in Colombia alone, and a great many more are probably still unknown. By far the greatest number of these birds live in the jungles, mainly in the tree tops, which makes them difficult to observe. The mere abundance of bird voices, however, conveys some impression of the great numbers to be found there. The most impressive bird groups in this area are the parrots: various parrots, huge, colorful macaws, and many parakeets, in addition to the toucans, with their grotesque beaks, and numerous species of iridescent hum-mingbirds. Countless aquatic birds—jabirus, herons, roseate spoonbills, scarlet ibises, etc., congregate near bodies of water.

A surprisingly great variety of bird life exists even in the most inhospitable areas of South America; in the Andes and at the southern tip of the continent. Along the lakes, often at altitudes of more than 13,000 feet, live untold numbers of ducks, rails, grebes, ibises, and flamingos. In addition to the various types of penguins that brood along the coasts and on the islands of the continent's southern tip, there are albatrosses, gulls, terns, ducks, and storm petrels, to name just a few species. The west coast and its offshore islands accommodate an extremely large bird population. The coast of Peru, for example, is inhabited by more birds than any other coastline on earth: one million cormorants, brown pelicans, boobies, gulls, Inca terns, and Peruvian penguins breed on the Guano Islands of Southern Peru alone.

The Galapagos Islands, about 620 miles west of the coast of Ecuador, are of great ornithological interest. Many of the 89 bird species identified there breed only on those islands; e.g., the waved albatrosses, the Galapagos penguin, the flightless Galapagos cormorant, the swallow-tailed gull, the dusky gull, Galapagos dove, hawk, mockingbird, and the famous Darwin's finches.

AUSTRALIA. Very unusual birds live in Australia and its surrounding islands; the mallee fowl, for instance, which construct brood ovens of earth and leaves for their eggs, the bower bird, and, in New Zealand, the flightless kiwis and the takahes, large flightless rails which were long believed to be extinct. Along the coasts of New Zealand, Tasmania and Southern Australia, there are great colonies of such sea birds as penguins, albatrosses, shearwaters, and terns. Numerous parrot varieties, including the many different cockatoos, breed in the warmer areas, particularly in the jungles of Australia. New Zealand also has parrots, and Tasmania has at least six varieties of them. New Guinea abounds with unusual birds, particularly in the foggy jungles situated at higher altitudes, where the magnificent birds of paradise mate and brood.

ASIA. The continent's most densely populated breeding grounds are to be found in the South Asian archipelago, where 340 bird species breed in Java alone. Breeding birds abound in the Indonesian island province of Sulawes (formerly Celebes), where 84 species—mainly rails, fowl and pigeons—are unique to the region. South Asia is a pheasant paradise where numerous varieties vie for supremacy in the splendor of their plumage. Innumerable aquatic birds brood there, among them various herons, ducks, tree ducks, rails, the odd pheasant-tailed jacana, darters, and kingfishers. The swiftlets, which live in enormous troglodytic colonies whose saliva nests are partly edible, also have their habitat in South Asia.

Japan has several very interesting bird sanctuaries with breeding grounds for such extremely rare species as the Japanese crane, Steller's albatross, the ancient murrelet, the Japanese crested ibis, and several vanishing breeds of pigeon.

India long has been fertile terrain for flamingos, some of whose breeding colonies are extremely populous, and for waders and other aquatic birds. Rare fowl species and a multitude of small bird varieties breed in the Himalayas. The oldest Indian bird sanctuary was established in the vicinity of Madras in the eighteenth century.

ANTARCTICA. Innumerable sea birds of many varieties, among them albatrosses and emperor penguins, breed mainly along the fringes of the antarctic mainland.

There are many variations
of the mating game ;
the marriage customs
of birds
differ greatly,
but for the survival
of the species,
it always takes
two birds : the pair

Self-preservation and preservation of the species takes precedence over everything. The instincts serving these purposes are the strongest, the most urgent. When its time has come, the drive towards procreation is irresistible. It always takes two to accomplish this most essential of all tasks.

The gregariousness of most birds may deceive us. Almost invariably we encounter our feathered friends in groups, or even in flights, and some species lead so sociable a life that only very close observation of the densely populated colonies makes it possible to identify individual couples, even during the breeding season. Nevertheless, even the largest assemblies—such as are observed among the sea birds— consist, in the essence, of individual couples, whether they have only just become acquainted or have already paired off.

Usually, the marriage partners will keep faith with each other. In their encounters they show moving, often effusive tenderness and only dissolve this intimate bond when their offspring can function independently.

Copulation, the process of physically uniting, can lead to fertilization only if the "state of mind" of the two partners is in accord. Each bird seeks to impress his or her chosen one with certain courting gestures and enticing sounds and will attempt to commend his or her beauty and prowess to the other; thus the couple will put each other in the right mood. Even when the marriage is dissolved immediately after consummation, each partner then goes its own way—at least for a few minutes all of the attention, all of the temperament, and all of the passion of the two birds is devoted exclusively to each other, to one single representative of the opposite sex.

Birds may be confirmed loners, like woodpeckers, or sociable, like flamingos; they may keep their chosen partner for life, like geese, or they may be promiscuous Don Juans, like peacocks; yet, the fulfillment of their most significant task, the continuation of life, always requires just two birds: the pair.

27

BUILDING THE NEST

The birds clearly try to refute the law of gravity and rule over the air space. To realize such a daring plan, great flexibility and adaptability are required.

Their reptilian ancestors buried their eggs in the ground and left them to be hatched by the sun. The resulting high rate of loss among eggs and offspring was compensated for by a corresponding increase in the size of the broods. Of course, this primeval method functions only in temperate and warm regions, and fails entirely wherever flight is required. The weight of many eggs within the female's body periodically would make flight impossible. For this reason, birds rarely have more than one egg at a time in the oviduct. Hence, they must ovulate at intervals of one to two days. Consequently, the number of eggs is limited, but the growing offspring are more carefully protected. Moreover, in order to make it possible to settle in less temperate zones, parent birds have to provide their embryos with the protection of their own body heat.

ARDEA PURPUREA, *Linn.*

The purple heron builds its sedge nest within the protection of the reed thicket.

Although coot eggs are deposited at ground level, they are well concealed in the swamp vegetation. Reeds often are drawn over the eggs to protect them against detection by flying nest robbers, and the coloration of the eggs provides excellent camouflage.

Mammals generally solve these problems by retaining their young inside their bodies until they have reached a relatively advanced stage of development. They can afford the weight increase that occurs during the gestation period with very rare exceptions — because they do not have to fly.

But the birds have learned to conceal their eggs, to protect them from their enemies, and to keep them warm under their bodies. During their search for an improvement of this method, they "invented" nest construction, and the multiformity of the nests matches the diversity of the birds themselves; we need think only of the hummingbird and the ostrich, the penguin and the common swift.

Every imaginable sort of nest exists, from the hastily made trough in the ground to the most skillfully braided spherical shelter; from earth cavities, and stick aeries weighing tons, to piles of foliage measuring several yards in diameter.

No one knows how birds learned to build nests. Here, as in so many other evolutionary processes, we must rely on conjecture and imagination. Nonetheless, the development probably went something like this:

The simplest process was to lay the eggs on the ground. But since eggs are both delicious and nutritious, the need for protection from various predators must have arisen rather early. What could be done?

The bird could simply sit on its brood as soon as the first egg was laid. Such a crouched, motionless bird is almost indiscernible be-cause of its grayish-brown color-ation, and because it is odorless due to the absence of sweat glands. This simple method is still practiced by various species, but it has a decisive disadvantage: the young hatch at one- to two-day intervals, which, of course, can be fatal for ground breeders. Nowadays, most birds that nest on the ground and produce early fledglings do not begin to warm their eggs until all of them have been laid. Hence, the embryonic development will set in simul-taneously, enabling all the young to hatch within a few hours. On the other hand, the late nestlings, all protected by well-concealed or otherwise secure brooding sites, safely can start to incubate as soon as the first egg is laid.

Since the ground breeders were unable to sit on their eggs imme-diately, they had to learn to conceal them, perhaps under a tuft of grass, to cover them with vegetation before leaving them, or to provide them with camou-flage colors and markings even before depositing them.

Some species looked to earth cavities, rock crevices, hollow trees, steep cliff faces, and islands for protection. Others learned to dig subterranean brooding cham-

bers or to chisel them into tree
trunks.

A creature that can use its flying
ability to evade its enemies
naturally finds it convenient to
shelter its precious offspring at a
high altitude. But there are not
enough tree cavities and rock
crevices, and it is both difficult
and precarious to deposit eggs
on a branch. The birds, being as
adaptable as they are, began
therefore to build simple platforms of dead twigs into the
forks of branches. The more
dexterous among them gradually developed bowls, delicately
braided from all kinds of plant
fragments and provided with an
insulating padding of fibers, seed
tufts, small feathers, animal
hairs, etc. This type of nest construction culminated in true
works of art: spherical structures
which were intricately entwined
and fastened to the end of the
most delicate twigs, and which
had long, drooping access tubes
—masterpieces which command
our greatest admiration.

The beauty of birds' nests, the grandeur of these structures, which often are made with unbelievable skill, hardly lends itself to adequate photographic presentation. Perfect photographic reproduction is frustrated by the mere fact that these cradles for the young are nearly always concealed, protected from hostile looks (and from the camera) by foliage, by filigrees of branches, by caves and crevices, and, moreover, by a confusing interplay of light and shadow. The painter can omit unnecessary accessories, and thus ornithological artists of all periods have

Contrary to the wading bird's custom of breeding on the ground, the female of the green sandpiper prefers to lay her four eggs in the abandoned treetop nests of thrushes and jays.

The blackbird constructs its nest of brush, grass blades, moss, foliage, and lumps of earth and rotten wood.

The starling raises its offspring in the protection of hollow trees, abandoned woodpecker holes, nesting boxes, and other hiding places.

Bohemian waxwings brood in the northern regions of the Old and New World, preferably forming loose breeding colonies in coniferous forests.

Anyone looking at the slovenly hay nests of the house sparrows would scarcely think that such superb architects as the weaverbirds could be their nearest relatives.

Crossbills generally build their brush nests between January and March, when conifer seedlings are available as food for their offspring.

Bullfinch couples build their nests of thin, dry twigs among bushes and low trees, preferably in spruce. The nest is padded on the inside with small roots.

The elegant long-tailed tit prefers to build its pretty, thick-walled, spherical nest of moss and delicate plant fragments in the sturdy fork of a tree, but bushes are often used as well. Usually, lichens and spider webs are used in adapting the nest, which is lined with many small feathers, to its environment.

Like the blackbird, the fieldfare fills the chinks of its brush nest with loamy earth, later lining it with blades of grass.

The hoopoe hides its eggs in hollow trees or holes in masonry. In case of danger, the brooding female and the young will exude a horrible odor. In addition, the nestlings will squirt excrement at the presumptive enemy.

done justice to this fascinating architecture. John Gould, probably the greatest of these masters, created a gigantic work a hundred years ago. In 41 volumes, he published more than 3,000 of the most minutely detailed bird paintings, many of which included nests. The twelve lithographs on these pages were reproduced from his two books, *Birds of Britain* and *Birds of Europe*.

The song thrush plasters the inside of its nest with quick-setting loam. The eggs are deposited on this hard surface without any added padding.

Gray herons build large aeries of sticks, and prefer to locate them on high trees. Less frequently, they will build reed nests in reed banks, which are renovated and re-used again and again.

No animal
has been able to extend
the limits
of possible survival
as far
as the bird.
Unequaled
adaptability
enabled it to inhabit
even the most inhospitable
locations.

Wherever nature has offered a chance for survival, be it ever so small, some bird has discovered it.

The tropical jungle, whose multitude of small living spaces, of "ecological niches," remains unequaled, naturally offered the greatest possibilities for development. Its abundance of vegetation and small animal life permitted the development of the most varied bird species as well as of nutritional specialists and provided all sorts of nesting opportunities. In the jungles of Colombia alone, almost 1,550 bird species were able to develop. Most of these build their nests in the treetops, while many others use hollow trees, nest on the bare ground, or in earth cavities they dig themselves, among the reeds or even on the water itself.

Along with temperatures favorable for survival, the jungle offers plenty of food and all kinds of protection. But birds also live, build nests and reproduce under extremely unpropitious conditions, often in particularly large numbers (although invariably fewer species are represented).

Even the high mountain climate,

ordinarily so hostile to life, fails to discourage the birds. While in most areas breeding sites at altitudes of more than 13,000 feet normally are but sparsely occupied, there are exceptions. Thus, hundreds of thousands of aquatic birds nesting at altitudes of up to 15,000 feet can be found in the Andes, along the lakes of the high plateau. On some of these lakes, various ducks, grebes, and rails are as numerous as on the well-stocked ponds of zoos, and the swampy shores teem with herons, ibises, gulls, and plovers. Anyone who has ever seen the nesting site of an Andean grebe, a float-

ing nest of rotting plant fragments soaked with icy cold water, will find it difficult to imagine that eggs can develop in it.

But not all broods seem to be quite so resistant. Ducks, moor hens, and other rails build their nests on rush-grown hummocks and small islands if such are available. Those who settle on the shore must be prepared for a surprise nocturnal visit by a hungry fox. Horned coot couples have a unique method of coping with the problem of island scarcity. They pile up stones below the surface where the water is about forty inches deep until the pile rises above the water level. Their nests are then constructed on top of this artificial island which may be more than ten feet in diameter at the base and will consist of an estimated one and a half tons of piled-up rocks.

Other birds are found in these regions, however, belonging to species that are even less readily adaptable than aquatic birds. There are several types of hummingbirds, for example, which usually seek the sun. In the gloom of the jungle, they build warm nests of delicate plant fibers

Numerous species learned to live in the midst of villages and cities, to look for their food there and even to reproduce in inhabited houses. The barn swallows, for example, will build their nests in stables, barns, kitchens, and even in factories.

in rock crevices, instead of their usual bushes and treetops. The rock flicker, to cite another example, will dig deep earth cavities in the jungle, although normally its nest is situated in a tree trunk.

The waters of the polar regions, with their enormous fish population, offer ample food for resident birds. But to nest on these shores is frequently an almost insurmountable problem. The South Orkney Islands, for example, between South America and the Antarctic mainland, are almost completely covered with ice. For only a very short time in summer, a few lichen-covered areas will appear, which will then be used by some seventeen bird species to brood in great haste. Thousands of birds will brood in close proximity, some separated from each other by no more than a foot.

Every nesting opportunity must be exploited in these regions. Even the smallest rock projection, crevice, or rock ledge is occupied. When there is no available nesting material, the egg is deposited on bare stone, at times, even in places which are continuously inundated by melted ice.

Until a few years ago, the breeding sites of the gray gulls in Chile and Peru were searched for in vain. Then came the discovery that these birds reproduce deep within completely arid deserts, often more than sixty miles from their daily water supply.

Birds also nest in desert areas, again at the extreme limits of possible survival. Most of them subsist on unbelievably small quantities of liquid. Others, whose water requirements are not so well adjusted to existing conditions, will fly enormous distances each day, to find the necessary water. Thus, in contrast to most of their relatives, gray gulls will not nest near the water, but in the middle of the entirely arid saltpeter deserts of Peru and Chile, often more than sixty miles from the coast. Some sand grouse also must cover great distances daily between their breeding sites and the nearest water hole. One of the most astonishing examples of adaptability, however, is provided by those birds which live among urban humans. Whereas other species had millennia in which to adapt to new environments, they had to solve the same problem within a few decades. Many species were unable to manage this, but some actually were able to increase their numbers considerably.

Among Bushes

Shrubs—mainly solid bushes—the underbrush along the edges of woods, and hedges are extremely popular nesting sites. Various herons, especially the smaller species, brood in the tops or interiors of bushes. Many of those cuckoo species that do their own brooding like to build their nests in such locations, and many species of parasitic breeders prefer hosts that build their nests in bushes. A great many passerine birds and most mouse birds and humming-birds conceal their nests in underbrush or shrubbery. Even some gallinaceous birds have given up the ground breeding customs of their relatives to nest atop bushes.

On the Ground

All ratites, along with tinamous and sand grouse, and most gallinaceous birds and nightjars breed on the ground. Various birds of prey, parrots, owls, cuckoos, and numerous passerine birds will deposit their eggs on the bare ground, in scratched-out troughs or in ingenious ground nests.

In Earth Cavities

The kiwis or apteryxes, some ducks, a few wood-peckers, various rollers, kingfishers and bee-eaters, some passerines, and the trogons dig earth cavities for their offspring. Some owls, parrots, and passerine birds utilize natural ground cavities or the burrows of rabbits or other animals.

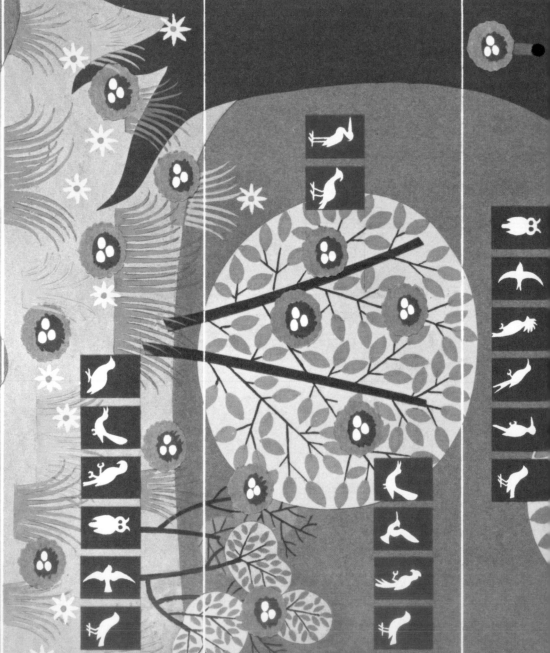

THE TREETOP BUILDERS

Most nests found high in the tops of trees are built by large birds. Herons, ibises, and storks usually form colonies there, while most birds of prey, owls, and crows prefer to nest by themselves. Passerine birds sometimes are found living as sub-tenants in large aeries that frequently are built up over many years, and some small birds attach their nests to the lower sections of the treetop aeries of birds of prey.

AMONG THE BRANCHES

Most pigeons and passerines, some parrots, cuckoos, swifts, hummingbirds, mouse birds, and a few nightjars build their nests among the branches of trees. These structures, which usually are bowl-shaped or spherical, may be braided and anchored to the branches, built in the crotches of limbs, glued to the sides of branches or to large leaves with saliva, or suspended from the outermost tips of twigs. Many of these nests are thoroughly camouflaged with lichens or moss.

IN HOLLOW TREES

Nearly all woodpeckers breed in hollow trees, and deserted woodpecker holes are often used by other troglodytic breeders, as well as passerine birds, rollers, and swifts. Parrots and trogons will enlarge hollows in rotten trees with their beaks. Various cuckoos, birds of prey, owls, and water fowl also lay

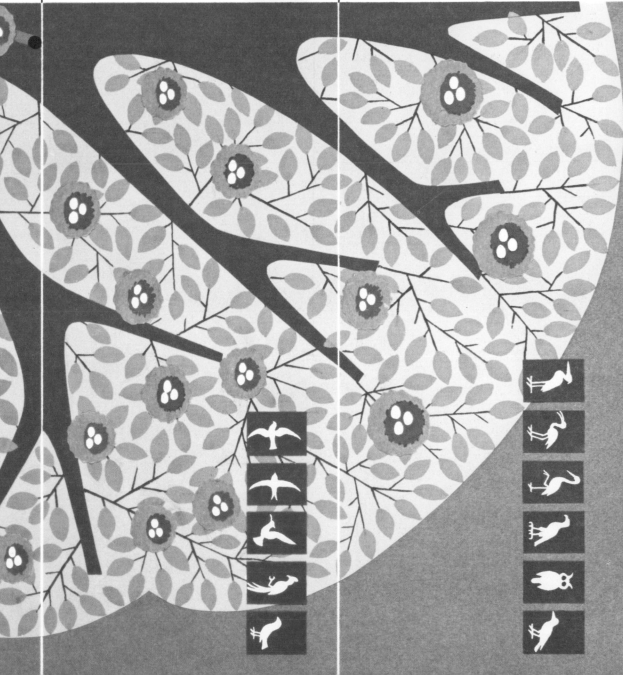

Among Waterside Bushes

Various boobies, pelicans, and cormorants, herons, a few ducks, gallinaceous birds and terns, some rails, and a great many passerine birds build their nests near the water within or atop bushes.

In Waterside Caves

Excavations and natural earth and rock cavities that are easily accessible to water are used by various auks, tubenoses and penguins, by a few ducks, and by most kingfishers.

On Beaches and Banks

Most loons, waders, gulls and terns, tubenoses, penguins, geese and ducks, various boobies and rails, as well as some owls, birds of prey, and passerines prefer to breed on the ground near the water. Many of them are satisfied with a simple earth trough, while others build nests of plant debris, shells, and small stones.

Among the Reeds

Some heron species and various ibises and spoonbills situate their breeding colonies in the protection of reed banks. Many ducks and rails, as well as some passerine birds, breed in such locations. Cuckoos often prefer reed-breeding passerines as foster parents for their offspring.

On Floating Nests and Rush-Grown Hummocks

Grebes, many geese and ducks, gulls, terns, and rails will build their nests on floats made of plant debris or on rush-grown hummocks.

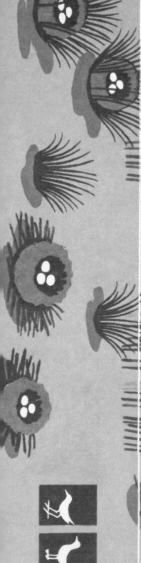
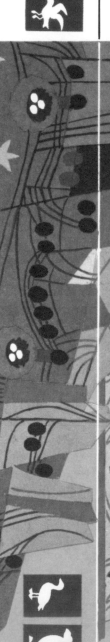

On Rock Plateaus

Various cormorants, gannets, tubenoses, auks, and some gulls breed on rocky plateaus, generally surrounded by sheer precipices and nearly inaccessible to their natural enemies. Almost invariably, they form crowded colonies, sometimes composed of various species.

On the Walls of Cliffs

Projections, ledges, niches, and crevices make excellent nesting sites on steep cliffs. These, of course, are well protected against enemies and are therefore preferred by various species. Some birds of prey build their aeries in large rock niches, which, as a rule, they will use again and again for many years, extending them annually and often making monstrous structures of them. Some owls and passerine birds will also brood here, isolated by couples, like diurnal birds of prey. Most of the other cliff breeders—various tubenoses, cormorants and gannets, auks, gulls, terns, swifts, and a few pigeons, nightjars, parrots, and passerine birds—will establish more or less crowded colonies in these same locations. For safety, the majority of these colonists will build extremely sturdy nests of seaweed, grass, or similar materials during the breeding season, and cement them to the rock with excrement. Swifts will mix the nesting materials with their viscous, sticky saliva, and in some cases may even make the nest bowls entirely of saliva, attaching these odd cradles directly to the sheer, precipitous wall of the rock. Auks, on the other hand, will deposit their pear-shaped eggs on bare rock ledges or terraces.

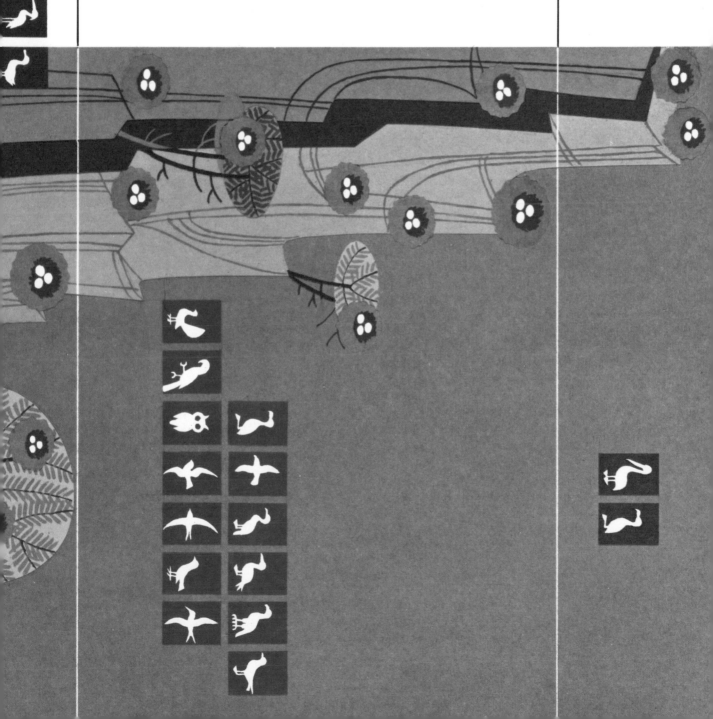

The protection of eggs and off-spring is indispensable to the survival of species. Birds have taken numerous approaches to the solution of this problem. One of the most effective of these was discovered by the "cave dwellers." Sand martins build their "pueblos" into steep banks of ravines and walls of gravel and clay pits in the Old and New World. It is hardly believable that these delicate, weak-footed birds are capable of excavating horizontal tunnels up to five feet long into the often surprisingly hard ground, and then enlarging the ends into brooding chambers. Hundreds—even thousands—of such nesting cavities may be located close to one another in favorable locations. Sand veins are preferred as nesting sites in gravel walls, as can be seen in the upper photograph of a sand martin colony.

The red-throated bee eaters shown below are gregarious, as are all their relatives, and nest in close proximity, taking up a minimum of space. Kingfishers, on the other hand, will dig their earth cavities sporadically, because every couple lays exclusive claim to a certain area for fishing purposes.

MASTERPIECES

Numerous birds—mainly among the smaller species—have achieved true mastery in nest construction. They braid, weave, glue, pot, and even sew in the process of putting together their little architectural marvels. Some of the most amazing of the cradles birds construct for their young are shown on these pages.

Sociable weavers use blades of grass to build enormous communal nests on sturdy tree limbs. The nests are used for years, enlarged annually, and eventually may be ten feet wide and sixteen and a half feet long. Up to a hundred and twenty family cells are built into the lower portion of these nests.

Scissor-tailed swifts use their sticky saliva to build tubes that dangle from overhanging rocks. These are about twenty inches long, have an entrance at the bottom, and a small platform for the eggs on the inner wall near the top.

Center right: Reed warblers build deep, delicately interlaced bowls attached to several reed stalks. These ingenious pile dwellings are usually located on marshy ground or above shallow water.

Right: Tailorbirds perforate the edges of one or two leaves, and sew them together with plant fibers or animal hairs. The softly padded nest is then built into this ingeniously constructed cone.

The South American Rufous Hornero builds spherical structures of clay on limbs and the crossbars of telephone poles. These harden rapidly in the sun and provide excellent protection for the offspring.

Tree swifts glue a tiny bowl to the side of a limb. This provides precisely enough room for the single egg, which is glued down with saliva for safety's sake. In relation to the size of the bird, this is the smallest nest known.

The mallee cock scrapes out a pit, about forty inches deep and six and a half feet wide, fills it with foliage, and then covers it with a sand pile approximately forty inches high and sixteen feet across. The buried eggs are hatched by the fermentation heat of the decomposing foliage—heat which is continually checked and regulated by the cock.

Some weaverbirds have achieved the ultimate in the art of nest construction. Their spherical nests are artfully woven of delicate stalks and plant fibers.

Horned coots will pile up stones under water to make artificial islands which may have diameters of up to thirteen feet at the base, may attain heights of forty inches, and which may weigh more than a ton. The nest of scraps of vegetation is then built on the dry tops of these islands.

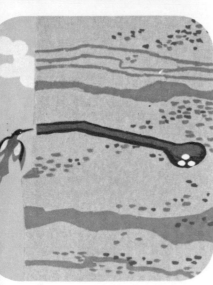

Bee eaters prefer to build their tubeshaped nests into the steep faces of loam banks or ravines. The cavities have an angular bend, are five to ten feet deep, and end into brooding chambers.

Most woodpeckers chisel pear-shaped cavities into tree trunks with powerful, whirling blows of their strong, pointed beaks. A few species dig brooding tubes into the ground or into termite hills.

Palm swifts use saliva to glue small, flat bowls about as commodious as teaspoons to palm fronds; saliva also is used to fasten vertically positioned eggs inside the bowls.

Bird Nurseries

◄ *Penduline tits prefer to attach*
their exceedingly ingenious nests
to the extreme ends of willow branches.
The nests are of thick-walled construction
and so solidly and warmly built
that children in Eastern Europe
collect them in fall
to be worn as slippers in winter.

The Foot as Cradle...

The two largest penguins—the king penguin and the emperor penguin—are the only birds that require neither nest nor established nesting site. They brood on the snow and ice of the Antarctic Circle. The solitary egg rests on the fat-padded feet of the mother or father and is warmed by an enveloping abdominal fold.

... and Rocks as Nesting Material

Some small penguins choose caves and rocky recesses for brooding, while others prefer open areas where their view is unobstructed. Some penguins dispense with nesting materials altogether; many others collect objects in the vicinity —stones as a rule—with which they build small mounds. Adelie penguins make something of a game of stealing the stones which their neighbors have managed to pile up.

The Fortress

Hornbills seek out hollow trees for brooding. The birds range in size between that of crows and geese, and the cavities must be spacious enough to accommodate them comfortably—and cavities more than twenty inches in diameter have been discovered. Once the nesting site has been selected, the female fetches small lumps of damp clay and walls up the opening until she barely can manage to slip through, compacting the mortar with vibrating motions and lateral strokes of her beak. More clay is then brought by the male and used by the female in further closing off the brooding cavity, leaving only a tiny cleft, through which she is fed, and refuse is ejected. The male sees to it that the outside is neatly plastered. When the wall has dried, it is amazingly hard and offers

excellent protection. Then, when the young have hatched, the female breaks through the wall and leaves the cavity, which is often walled up again with fresh clay immediately by the young birds for their own protection.

They Sacrifice Plumage to Feather Their Nests

Female ducks will pluck their own delicate abdominal feathers for lining their nests (usually somewhat hastily constructed on the ground) and for camouflaging their eggs. Eider ducks produce particularly soft, abundant down feathers. (Eider down has become world-famous as filling for quilts, cushions, and sleeping bags, and, much to the distress of the ducks, is frequently taken from nests in which the young have not yet hatched.) Brooding is done by the female alone, who goes out to sea at night in search of food. Were she to leave the nest without being able to cover the eggs with a thick lining of down feathers, they would cool off rapidly and the embryonic chicks would die in the chill northern nights.

Tailoring

The nest of the tailor bird of South Asia is among the most astonishing of all bird structures. These masters of the art of weaving perforate the lateral edges of one or more leaves with the point of their beaks. Then they look for wool, plant fibers, or even spiders' webs, sew the leaves together into a conical roll and construct a bowl-shaped nest of delicate fibers and animal hair within the cone. Some members of the warbler family—to which the tailor bird belongs—also attempt to join perforated leaves by sewing, but the results are clumsy by comparison.

The Saliva Nest

The edible "bird's" nests offered as a delicacy by Chinese restaurants, are world-famous. Actually, they are not produced by swallows, as is commonly supposed, but by swiftlets, which are related to goatsuckers and hummingbirds. For a long time, the materials used in construction of these avian cradles remained a mystery. It was assumed that they came from the sea and were collected along the shore, since these birds brood in large caves along rocky coastlines. It is now known, however, that swiftlets make their nests of their own saliva. Before the breeding season, the salivary glands of these birds swell considerably and produce a viscous, colorless mucous secretion. With lateral movements of the head, the birds deposit this secretion, layer by layer, on the precipitous rock wall to form a bowl-shaped nest which soon takes on a horny consistency. Nest collectors reap their harvest while it is still as fresh as possible. The bird starts building again as soon as the nest has been removed, and within a single breeding season, an adept collector can rob the architect of its nursery three times before the frustrated bird is finally permitted to brood and raise its young in peace. An efficient collector can reap almost 40,000 nests within two months, a harvest that will net him from 3,000 to 4,000 dollars—provided he does not fall off the long, swaying bamboo ladders he uses in his dangerous work.

Earth Caves

Bee-eaters perform difficult earth-moving feats in the construction of their nests. Their horizontal tubular tunnels, usually situated in a sharp bend with an enlarged brooding chamber at the far end, are dug into sloping banks or clay and sand walls, the birds making use both of their beaks and short feet. On the rarer occasions when their caves are dug into level ground, they lead diagonally down-ward. The length of the passage usually varies from three to seven feet, but in exceptional cases may be up to ten feet long. The snowy white eggs are deposited on the floor of the brooding chamber while it is still bare. Gradually, the chamber is neatly furnished with a nesting material consisting of indigestible insect particles regurgitated by the birds. As many as 2,753 insect heads have been counted in a single bee-eater nest.

Leased Lodgings

Some larger bird species will use the same nest year after year. An even greater number, however, tend to move into old abandoned nests—often built by other species—whose size and location appeal to them. Thus, a buzzard's nest may be occupied first by the original architect, a year later by a pair of brooding tawny owls, and, the following summer, by a long-eared owl who might raise its young there. Soon after the nest is occupied, and usually throughout the brooding and rearing period, it is renovated, enlarged, or converted—a process which may turn it into a huge pile of brush over the course of several years.

Multiple Family Dwellings

Probably the most conspicuous birds' nests of all are those of the weavers. These gregarious, drab-colored grain eaters, approximately the size of sparrows, build a thick rain-proof communal roof of grass blades, reminiscent of the thatch on a native hut, on a sturdy tree limb. Each couple will then install its own nesting chamber under this roof. As a rule, such a community nest will be continually enlarged and used for many years, usually until the supporting branch breaks under its weight. Nest diameters normally vary from six to ten feet, but some with diameters of more than

sixteen feet, fitted with more than 120 individual dwellings, occasionally have been found. Unoccupied apartments frequently will be used as brooding sites by such other birds, as small parrots and even falcons.

Similar nest constructions are found among an entirely different order of birds—parrots. Gray-breasted parakeet, for example, build large piles of brush in tree tops, where each couple will install an individual cell. They, too, will use these nests for many years, not only for breeding and raising their young, but also as sleeping quarters before and after the breeding season—a very rare custom among birds.

Throw-away Structures

Small birds almost invariably leave their nests after a single breeding season and build anew the next. This continual apparently superfluous rebuilding is actually quite necessary, however, for, at the end of a breeding season, the structures are usually infested with numerous mites and other parasites which might seriously endanger the next brood of hatching young.

Timbered Nests

It is well known that woodpeckers make their breeding cavities in tree trunks. They are usually pear-shaped, with their entrances near the top of the cavity. What is not so well known is that these cavities are chiseled not only into rotten trees, but also into healthy hard wood. This is a difficult task, requiring a beak equipped with an extremely hard, horny sheath and a special type of edge, and terminating in a sharpened, chisel-shaped point. The bird is also equipped with a reinforcing bone structure at the root of the beak and powerful neck and head muscles which lend force to the hard blows on the one hand, and, on the other, help to absorb them, thereby protecting the brain from damage.

Not all woodpeckers, however, brood in hollow trees. The Gila woodpecker of North and Central America often builds its cavities in enormous cactuses that are not as hard as trees, but whose spines offer a different kind of protection. Various American woodcreepers dig earth caves into rotting tree stumps at level or sloping ground.

THE FLOATING NESTS

Grebes' nests actually are rafts constructed from all sorts of plant fragments. Sometimes, these rafts float freely, but usually they are anchored to reeds, rushes, or other plants along the water's edge. The birds build up considerable piles, which soon become waterlogged, begin to rot, and gradually sink. By the end of the brooding period, only a small portion of the nest remains above water, and the young are hatched in a rather damp cradle.

THE EGG ON THE LIMB

Terns do not bother much with nest construction. A slight recess in sandy ground, sometimes decorated with a few small shells, stones, or bits of vegetation, generally serves their purposes. Only a few species build tree nests—in most cases slightly concave brush platforms in bushes and low trees. The fairy tern—when it does not retire to rocky promontories—seeks refuge in the height of tree tops, but it does not build a nest: it deposits its eggs on a level part of a branch, a spot that is often so small that the bird is unable to sit properly and is forced to brood in a half-squatting position.

DORMITORY NESTS AND OTHER MOCK NESTS

The males of some wren species will build several nests in their area. Usually, they are spherical moss structures with an entrance on the side. A single wren will have four or five nests and will use them alternately for sleeping quarters throughout the year. He will rarely make repairs when one of the nests is damaged or begins to disintegrate, but prefers instead to build a new sleeping quarter. At mating time, the male wren will show one of his nests to his bride-elect, and, if they marry, she will carefully provide it with a lining.

Some wrens, particularly younger birds who have left the parental nest, like to spend the nights companionably during cold weather and will snuggle up closely to one another during the nights. As many as twelve wrens have been counted in one such dormitory nest.

Various birds, loons for instance, build nests whose purpose is questionable, but which may serve to decoy predators away from inhabited sites.

THE SLIPPER NEST AND NESTS WITH SAFETY LOCK

Many of the handsome and vivacious members of the tit family have been eminently successful in adjusting to man-made alterations of the landscape. The true tits are troglodytic breeders and build their warm nests in hollow trees and nesting boxes. Many tits, incidentally, also sleep in hollows outside the breeding season, and often sleep alone. In spring, the nest of the female usually is employed as the nursery. Natural breeding cavities, however, are growing increasingly scarce these days, as sick, hollow trees are cut down, and if there are not enough of these natural nesting places, the tits must look for other sites. Although they are capable of building nests among the branches, they do so very rarely. They prefer to look for emergency quarters, which often have a tendency to be quite unorthodox. Tit nests have been found in mail boxes, lanterns, watering cans, canisters, venetian blinds, behind window shutters and in drain pipes. One blue tit even had its nest in a sluice, specifically in the cast-iron housing surrounding the screw mechanism for raising the flood gate. The entrance was the hole through which the crank for lifting the sluice board was inserted.

Several representatives of the tit family are not troglodytes, and some build very ingenious nests. The nests of the penduline tits, for example, are masterpieces. As a rule, they are built on a drooping branch, preferably of a willow or poplar. The male penduline tit begins by braiding a loop of long plant fibers, 10 inches high and $4^1|_2$ inches wide, which hangs down vertically. Then he widens and completes the bottom end of the loop to form a bowl-shaped structure, looking, on the whole, like a basket with a handle. Next, he closes the rear and builds up the front and the sides until there is only a small opening left at the top, which he finishes off by extending into a tubular appendix that serves as the nest's entrance. During the final phases of the work, particularly in constructing the lining, he is assisted by the partner he has courted while construction was under way. The nest swings in the wind like a pendulum (which accounts for the bird's scientific name Remiz pendulinus) and is very sturdy, with thick walls interwoven with soft seed fluff and similar insulating materials. In Eastern Europe such nests are frequently collected in autumn by children who use them as slippers.

The nest of the African penduline tit is even more peculiar. It has a mock opening in the form of a sacklike cavity below the actual entrance. The true entrance is squeezed shut by the adult birds on leaving the nest. This undoubtedly misleads many a nest robber, who investigates the mock cavity only to find it empty.

Birds that feed their children
crawling insects, spiders, worms,
and similar small creatures
usually build their nests
at a great distance from each other,
on their own jealously guarded precincts.
In cases where the young are fed
with vegetable substances,
flying insects or fish, the birds—
which are almost invariably gregarious
by nature—nest and brood in colonies,
since there is little competition for food
in the immediate vicinity of the commune.
The photograph shows a colony
of grain-eating weaver birds.

GIANT NESTS

Storks generally brood in the same nest year after year. Each spring, before laying their eggs, they carry new branches and twigs to the nest, which is then lined with hay and, on occasion, even with stolen bits of fabric. Over the years, this brush aerie will grow into a structure that may weigh well over a ton. The aeries of many birds of prey are enlarged in the same fashion and may even grow larger than the castles of storks. In Florida, bald eagles built a nest which was almost ten feet wide, more than twenty feet high and weighed an estimated 5,500 pounds.

NESTS NEAR A WATERFALL

The dippers are relatives of the wrens. They have absolutely unique habits among songbirds: they live along brooks and streams, and they search for food mainly by diving into the water and running about the bottom. The nesting sites of these birds also are unique. The nests are built of damp moss and grass, and often are hidden behind a waterfall. Frequently, the birds can reach them only by flying through the cascading sheets of water.

GLUED MINI AND MAXI NESTS

Most swifts employ conventional materials in nest construction, such as leaf fragments, animal hairs and feathers, but they glue them together firmly with a saliva that dries quickly and turns quite hard. The reason for this gluing of the nesting materials will be clear to anyone who has ever seen a swift caught on a loose stalk or thread. Once its claws become entangled, the swift frees itself only with difficulty, for its feet are very short and its bill poorly suited to the task. It thus happens occasionally that one of the birds is inextricably caught and starves to death; a fate to which similar species are not immune.

The nests of most swifts are bowl-shaped and about the size of a man's palm. They are stuck to the inside walls of rocky caves, hollow trees and buildings, high above the ground. But some species do not follow this general trend. Tree swifts will glue a bowl consisting of well-cemented bark particles and small feathers to the side of a limb. In this case, however, the bowl is smaller than a teaspoon, and, relative to the size of the bird, is probably the smallest of all nests. These birds lay only a single egg which they glue to the mini nest for safety's sake. The nest looks like a small bulge on the limb when seen from below.

Still more peculiar is the nest of the Old-World palm swift. It is also a tiny, flat bowl which adheres to a drooping palm frond and is barely able to hold the two eggs that are glued down, in a vertical position, inside it.

The nests of the lesser swallow-tailed swifts are monstrous structures by comparison. They build irregular tubes which are twenty to twenty-four inches long, look like coat sleeves, and dangle from overhanging rocks or from the lower sides of tree branches. These tubes are open at the bottom and have a small concave platform for the eggs on the interior wall near the upper end.

THE MUD NEST

Muddy banks, salt lake islands and, occasionally, shallow waters serve as flamingo breeding grounds. Here, these birds, often gathered in enormous flocks, build their most peculiar nests. They pile up the mud directly beneath their bodies with their hooked beaks, tamping it down repeatedly with their webbed feet. Stones, shells, and other objects that can be reached from the nesting site with their beaks, are also included in these constructions. The truncated mud cones produced in this fashion dry in the sun and harden. They are usually about twelve to eighteen inches high and have a shallow recess for the eggs at their tops. On exceptional occasions, flamingos will also brood on hard, rocky ground, and in such cases they will dispense with nest construction altogether.

PILE DWELLINGS

The nests of most reed warblers are little masterpieces of ingenuity. The well-braided, deep, bowl-shaped nests are anchored to reed stalks, from two to five in number. The rough surface of these stalks prevents the airy pile dwellings from slipping down into the water below.

THE HAMMOCK NEST

The female of the golden oriole, which is about as large as a blackbird, builds her nest high in a tree top. She uses saliva to fasten a number of plant fibers and grass blades to the fork of a branch, usually at a considerable distance from the trunk, and then deftly weaves similar stuffs into a basket that swings between the branches like a hammock.

COURTING CUSTOMS

When the mating season approaches, the attitude of birds—their habits, appearance, and behavior—changes to an astonishing degree.

In the autumn, certain species will come together in great flocks, and move about harmoniously to reconnoiter favorable feeding grounds. As the mating season draws near, however, they become more and more quarrelsome and attack each other ever more frequently. Gradually, the flocks thin out, they break up and scatter in all directions, and the individual birds seek a suitable place for mating, a site which they will defend vigorously against every intruder of their own kind.

Some birds, on the other hand, will only begin to become sociable at mating time and will form breeding settlements, colonies where hundreds or even thousands of birds will brood in close proximity.

Many species pursue a strictly monogamous existence, either for a single season or for life, while others practice various forms of polygamy.

In each and every case, however,

certain preparations must be made, without which mating would be impossible. These ardent, stimulating mating preparations are generally known as courting.

During courtship, birds display an awesome variety of colors, movements and forms—a variety that is matched by the multiformity of their nests and the variation

The methods used by male birds to sue for the favor of the females are almost boundless. At mating time, the male frigate bird inflates his usually almost invisible throat pouch into a brilliant red balloon as large as a child's head.

of their general appearances. Many of the courting males don a resplendent plumage; they dress up, so to speak, and then parade themselves, almost in "dandy fashion," to impress the females, which are usually plain but, characteristically, choosy. In addition, they frequently present themselves, singly or together with other suitors, in ecstatic dances, sometimes performed directly before the bride-elect and sometimes at a considerable distance away, apparently with utter disregard for the other sex.

Males of the drabber species will attempt to compensate for their lack of feathered splendor by every conceivable sort of caper. They will perform wild aerial acrobatics including spirited loops and rolls. They will perform grotesque dances. They will sound drum rolls on hollow objects with their bills. Some will build the framework of a nest and commend it to the lady of their choice. And there are many whose most attractive qualities lie in their throats. They will entice the object of their desire into their home grounds with warbled stanzas, tender whispers, gay trills, or mellow sighs.

DRESSING UP
FOR THE WEDDING

Amid
the deadly frosts
of winter,
birds discard
their drab colors,
and their feathers
become
magical
with resplendent colors
and brilliant hues—
this is
the miracle
that announces
the approach
of spring.

*Although the male Lapland bunting does
not develop a resplendent plumage of brilliant colors, he nevertheless manages to spruce
up a bit when on the lookout for a mate.*

Birds sometimes start their mating preparations as early as autumn, but usually they do not begin to get "dressed up" for the occasion until the middle of winter.

Not all birds undergo a striking metamorphosis during the mating season. In some species the normal plumage and the mating plumage differ only slightly; the colorful dresses often fade almost imperceptively after the breeding season, losing but little of their brilliancy. Some species, however, are almost beyond recognition when they have donned their courting costumes. There are weaverbirds, for instance, whose ordinary plumage does not surpass that of a sparrow in any way,

head plumage at mating time. The male ruff, whose plain, light grayish-brown autumnal dress is identical to that of his mate, wears a unique costume for the mating game: around his neck he sports an enormous feathery frill which he can erect in a resplendent array of colors, which may include white, rusty red, chestnut, or black and a variety of intermediate shades.

It is not only the plumage which may become more resplendent at mating time. The bill or the bald areas on the head can also change conspicuously. Wartlike excrescences may appear, particularly on the head, or they may increase to many times their original size and change their color.

The plumage of birds performs many more functions than the skin covering of most other animals. It guards the body against cold and heat and protects the skin against injury. It provides the angular figure of the bird with its marvelous aerodynamic lines. It enables the bird to attain unequaled flight performance. It can camouflage the bird just as effectively as it can adorn him. Individual feathers are used by some birds as sound generators, as intermediary tactile organs, or as nesting materials. The owl's

other birds except possibly the kites or gledes have dispensed with.

Without its feathers the bird would be an unprepossessing, helpless, doomed creature.
Feathers, which have evolved over the eons from reptilian scales, are, even in their present stage of development, of a rather horny construction, devoid of sensory nerves, and attached loosely to the skin.

Despite the infinite variations in the size and shape of feathers we differentiate among four main feather types: 1) the contour feathers, which are of the greatest importance for the bird, providing its external form and protect-

Black-necked grebe. Grebes of both sexes wear a magnificent wedding costume.

The great northern diver undergoes a particularly conspicuous transformation at mating time.

but for courtship they display themselves in blood-red splendor. The usually white-headed laughing gull dons a chestnut-colored

feather ruff, which is responsible for its round facial features, acts as a kind of hearing trumpet—an outer ear or auricle which all

ing it against weather and injuries; 2) the large, sturdy flight feathers (known as pinion feathers on the wings and as rectrices

on the tail), which render the bird volitant; 3) the soft, fluffy down feathers which function almost exclusively as heat insulators and are located beneath the envelope of the contour feathers; and 4) the simply constructed filament feathers which mainly surround the base of the bill and the eyes of some birds; although themselves nerveless, may transmit tactile sensations to the flesh in which they are rooted.

Despite its enormous durability, the plumage of the bird is naturally subject to wear. It must be removed from time to time. In contrast to the hairs on our head, which grow continuously, the feather of a bird attains a certain size and then ceases to grow, even if it is cut or broken off. Every bird must therefore change its plumage periodically. This process is known as "moulting" and occurs over various time periods. With most birds it lasts from four to six weeks, but, as in the case of the eagle, for example, it may extend over an entire year.

With most birds, the plumage of the male plays a particularly important part in courtship and mating, even when he does not change to a more resplendent

dress for his wedding. Each in his own way will seek to display the beauty of his feathers to best advantage in order to impress the female. He will ruffle his feathers, spread his wings, or let them droop down. These plumage displays are frequently accompanied by the oddest of gestures and body contortions.

Male pigeons will draw themselves up to their full height and circle their females with mincing

steps, showing off their iridescent or conspicuously marked necks, and some species will spread their rectrices, exhibiting a handsome fan. Lapwings, hoopoes, and many other birds will raise the plumed crests on their heads. The male great bustard with his plain camouflage coloring will lay his rectrices along his back, revealing soft white feathers, normally hidden beneath his tail,

which will cover his back like a light blanket of snow. Herons will ruffle the delicate, bushy ornamental feathers on their backs. Golden pheasants have exceptionally colorful feather ruffs which they raise at mating time, spreading them forward until they cover their bills. Ringnecked pheasants dazzle their hens with the colorful iridescent splendor of their wings by posing sideways in front of them and spreading the wing on the side facing them all the way down to the ground. The delicate black and white feathers of the male ostrich, primarily those on the wings, play an important part in his expressive courting ritual. He erects his wing feathers as impressively as possible, while he alternately spreads his wings wide and beats them, or lowers them, trembling, toward the ground. At the start of the courting ritual, the male golden-headed manakin turns his backside on the female, raises his tail, and displays his brilliant red-plumed shanks.

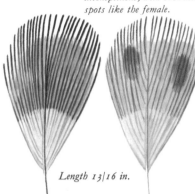

Head feather of the great crested grebe. Male and female grow a great erectile ruff of beautiful rust colored feathers before mating.

Head feather of the male northern mallard. At mating time, the head and neck are an iridescent metallic green. At other times, they are inconspicuous, with brownish spots like the female.

Head feather of the brambling. The male's brownish spotted head and back turn brilliant black at mating time.

Head feather of the black-headed gull. The head plumage of both sexes is chocolate brown at mating time. At other times it is white with a dark spot in the auricular region.

Length 13|16 in.

Length 5|16 in.

Length 9|16 in.

◄ *The white wagtail. Among small birds, there is generally less difference between the plumages in the mating season and at other times.*

The special sexual finery of the dunlin consists of a black abdominal shield. ►

The European spoonbill—male and female alike—grows a magnificent tuft of feathers at the back of its head, and the base of its neck turns a yellow ocher, in order to be especially beautiful at mating time.

Concealed
in the mists of the jungle
he dances
for his beloved:
the most magnificent
of them all.

"Birds of Paradise" was the name which the Dutchman Jan van Linschoten gave to the rightful owners of the fantastic bird skins which sixteenth-century sailors first brought to Europe from New Guinea. The birds were skinned by the natives to conserve the magnificent plumage that was still unknown elsewhere in those days. For this reason, Europeans firmly believed for three hundred years that these birds had neither feet nor entrails and hence, like the inhabitants of paradise, were not subject to decomposition. It was claimed that these animals spent their entire lifetime in flight, that they subsisted exclusively on the dew of the heavens, and that they even did their breeding in the air, the female sitting on the back of the flying male.

While these birds do not come from paradise, as it was later learned, they seem to be fully justified in bearing their name, nonetheless: their beauty is truly heavenly. The magnificent colors and the multiformity of plumage among male birds of paradise borders on the fantastic.

A male bird which dons a particularly conspicuous plumage generally does not concern himself with the care of the brood. This he leaves to the plainly-colored female. This also holds true for birds of paradise.

Usually in the early morning and often again towards evening, a number of these resplendent birds will assemble to do their courting as a group in the tree tops or, in the case of some varieties, on the ground of clearings in the misty jungles of New Guinea. Females will also put in an appearance, but they merely sit about, somewhat prudishly and unconcerned, as if their presence were a mere coincidence. Nor do the males appear to pay the slightest attention to the females, although the entire lavish display of colors, forms, and motions is staged solely for their benefit.

Each variety has its own peculiar courting rituals. The greater bird of paradise will incline forward and lift his wings with a shrug, raising the dense profusion of his long, delicate golden yellow back feathers, his whole body vibrating as if in utter ecstasy. He will hop and flutter from limb to limb constantly displaying his flowing feathers. The Prince Rudolph's blue bird of paradise will initially sit on a limb, sounding his calls of enticement, then he will bend forward until he hangs from the limb upside down. In this position, he will suddenly spread his wings and unfold a flowing profusion of rose-colored and iridescent blue ornamental feathers. The lesser superb bird of paradise erects his mighty black-feathered crest and opens the brilliant blue ruff on his neck, making him appear more than twice his size. The magnificent bird of paradise spreads his great ruff like a huge umbrella, hiding his entire body from the neck down. Only the shimmering red and golden throat and the long, stiff threadlike plumes with streamer-shaped ends at the back of the head appear above the outspread ruff.

Today, nearly forty bird of paradise varieties have been identified by scientists. The smallest of them are sparrow-sized, while the largest are as big as crows. They are probably close relatives of the chough and the Alpine chough. Even about the most familiar types of birds of paradise we do not know a great deal. Many varieties have not been investigated as yet; some may not even have been discovered. The wild, inaccessible mountain jungles of New Guinea will guard many of their secrets for a long time to come.

BIRDS OF PARADISE

Birds of paradise live for the most part in the almost inaccessible misty mountain jungles of New Guinea. This makes it nearly impossible to photograph them in their natural environment in such a way that the great delicacy of their plumage and the stunning splendor of their colors can be shown to best advantage. For this reason, mounted specimens were used for the photographs on the following pages.

Bottom:
The king bird of paradise is smaller than a starling, making him the smallest of all birds of paradise; but his beauty is in no way inferior to that of his relatives. When he is seized by the mating urge and wants to capture the attention of the plain female, he spreads the feather tufts at both sides of his breast to form fans with iridescent emerald green edges, and erects the two thread-like middle feathers of his tail with their rolled, medallion-shaped ends.

On pages 58/59:
During the courtship ritual, the blue bird of paradise, hanging head down from a branch, unfolds an abundance of magnificent ornamental feathers, which are predominantly blue, violet, and, on occasion, partly rose-colored. Meanwhile, he swings back and forth in time to his strangely hoarse calls.

On page 60:
While six-plumed birds of paradise are less splendidly colored than other species, they compensate for this shortcoming with a more eccentric plumage and with their great talent for grotesque dancing displays.
During the courtship ritual they will remove fallen leaves and other objects from an area of rising ground, and there display themselves in the most capricious manner.
Spreading their long neck feathers like an umbrella and hopping back and forth and in a circle, they look like black mushrooms with birds' heads.

THE BOWERBIRDS AND THEIR GARDENS OF LOVE

No less fascinating than the true birds of paradise are their closest relatives, the sub-genus, the bowerbirds. They are less ostentatious in appearance (although the males of most species wear a handsome plumage of black, yellow, and blue hues, and they often have an erectile golden feather crest as well), but the manner by which they seek to gain the favors of their plain-looking female partners is quite unique among the members of the feathered tribe: they build gardens, lovers' bowers, and ornamental turrets. While there are highly talented architects among other bird species, their construction activities are limited to nests for their offspring. The structures of the male bowerbirds, on the other hand, are not intended for depositing eggs; the actual nest is always located elsewhere among high trees, and is usually built by the female alone. The lovers' retreat is to the bowerbird what the splendid plumage is to the bird of paradise—both play a unique role in the mating game. Each male bowerbird builds his own dance floor for the courtship, clearing a piece of ground in the forest, from three to five feet in diameter. Some species leave this area bare, others cover it with plucked ferns or lichens of a color that contrasts with the immediate environment, and still others surround the whole area with a wall of moss and brush. The site is then decorated with colored berries, snail shells, iridescent wing cases of beetles, flowers, and other colorful objects. For some species this will suffice, but for most, it is only the beginning. According to the structures they build, bowerbirds are divided into several groups: the "maypole builders" will pile up small branches around a young tree in the center of the area, interlacing them cleverly to form a tower. Often they will drape this tower of twigs with moss or lichens and adorn it with the same colored objects which they used in decorating the floor of the area. The finished structure is generally three and a half to seven and a half feet high, and occasionally they are built up to ten feet high.

The structures of the "avenue builders" are even more impressive. These birds construct from two to four walls of intertwined branches, connected at the top. The ground in front of the entrances to these "avenues" or "bowers" is again usually lavishly decorated. Some of these bowers are actually small huts, about twenty inches high and three and a half feet in diameter, supported by a small tree trunk in the center, which is decorated like a colorful maypole. There are at least two species of bowerbirds who actually daub the walls of their huts with paint, which they manufacture from crushed berries and other vegetation, combined with charcoal and saliva. The paint is applied with the bill or with a small piece of bark held in the bill like a brush!

The lover's bower of this gardener and cabin builder may be forty inches high and five feet wide at the base. It is constructed around a young tree, which is used as center support. The bird— only as large as a blackbird—lays out colorful "flower beds" made up of varicolored blossoms, berries, lichens, etc., in his meticulously tidy front yard. He continually scrutinizes this little ornamental garden, immediately removing anything that has begun to wither and fade, and replacing it with fresh decorative material.

This marvelous courting area is used by the bowerbird for several months—sometimes even for a number of years—during which time he will continually putter about the place, making repairs, enlarging it, removing dead leaves and rotting berries and replacing them with fresh ones. He lures the female with a rather simple song, which he sings while standing inside the courting area or on a limb above it. When the longed-for visitor arrives, he will pose in front of his building or inside it. If he is successful in gaining her affection, she will enter his bower and the mating will take place in this "garden of love."

The frigate birds also have an odd courting ritual. When the mating season approaches, the bald spot on the male bird's throat begins to swell, and continues to inflate until it eventually turns into a brilliant red balloon as large as a child's head. Equipped for courtship with this startling appendage on his neck, the male goes in search of a suitable nesting site. Like most sea birds he does not object to company and will establish a loose colony with others of like intentions. By preference, he will choose the top of a bush as the site for the future nest but, if necessary, he may stake his claim on the bare ground or shore of an island.

When a female approaches, each male will do his utmost to win her for himself. Laying his head on his back to show off his throat sac, which is puffed up nearly to the bursting point, he rattles his bill, utters screeching calls, spreads his wings, tosses his body from side to side, and shakes himself so violently that the rustling of his feathers can be heard over a great distance. The female will often hover above the courting males for several minutes, as if to indicate that the spectacle pleases her, thus further exciting her contending suitors. When she at last alights near the male of her choice, the rejected candidates patiently wait for another to show up.

After the couple has come together in this fashion, the balloon lure is soon deflated, and nest construction begins. Now it is mainly the female who guards the nesting site selected by the male, while he goes in search of twigs for the construction of the simple nest platform. He almost never alights while breaking dry twigs off the bushes or collecting them from the ground. During this time, other males often will attempt to snatch the nesting material from him in a form of aerial hijacking that often involves prolonged pursuit, so that it may take quite a while until the couple has at last prepared its nest for the egg.

His crop puffed up for the courtship ritual, the male frigate bird, although still unattached, goes off in search of a nesting site. The unwieldy balloon impairs the otherwise perfect maneuverability of this most accomplished of all soaring birds.

Many species of sea birds brood in colonies. Undisturbed islands and quiet stretches of coast line offer little brooding space in comparison to the food supply provided by the oceans. Since these birds are not restricted to their immediate nesting territory for their food source, as are songbirds for example, but must fly further away for this purpose anyhow, there is no harm done by brooding close to each other. On the contrary, the community provides them with protection

frequent and noisy acoustical expression (particularly among the gulls), which certainly constitutes one means of communication, the language of gestures is probably far more expressive and versatile. When a bird wants to assert its right to a nesting site upon the approach of a new arrival, it will often raise itself to full height and spread its wings; in addition, it may open its beak aggressively accompanying these gestures with loud screeching. At this point the owner is almost in-

against nest robbers, since some of the parent birds on the breeding ground will always stand ready to drive off jointly any undesirable intruders.

But community life necessitates communication. While there is

variably acknowledged and appeased by the opponent, who will turn its head aside or will turn completely away in a cowed attitude. Its beak may be concealed in the breast plumage as a disarming gesture, or the closed beak may be held up vertically, symbolically demonstrating to the adversary that no attack is intended. These gestures are supplemented by numerous, more delicately shaded ones expressing affection, aversion, goodwill and humility, intended departure, readiness for mating, and many other things. Holding up nesting material with the beak to the partner is a widely accepted indication of the readiness to begin nest construction.

Feeding plays an important part in the mating rituals of terns. The male tern arrives with a small fish in his bill, seeking to win his lady by this means, as demonstrated by the least tern at left below. By all appearances he intends to convey the idea that he is an accomplished fisher and eminently capable of looking after a family. If the female takes a liking to him, she will accept the tidbit. At times, the male seems to regret his generosity at the last moment and swallows the fish himself. On the other hand, the female may indicate her willingness to mate by making begging gestures, even though the male has brought no food, as in the case of the arctic tern on the right.

The threatening gesture, which the kittiwake, at right above, backs up with loud cries, is answered with an appeasing gesture of humility by its vis-à-vis, which disarms itself by pressing its bill against its breast.

When a fulmar lands alongside his partner, he greets her by screeching with his head thrust forward. These birds seem to have less emphatic gestures than those of the gulls. But since they are typical birds of the high seas and excellent flyers, their feet are so weak that they are unable to stand on dry land.

Left: Despite the expressive gestural language of the masked boobies, they, like other gannets, frequently engage in beak duels which occasionally degenerate into bloody battles.

On the following pages: The drooping wings of the arctic tern mean: Would you like to marry me? Along with this gesture, there are a number of other motions indicating her desire to mate.

Avian Choreography

Song, splendid colors, and peculiar contortions are not the only enticements used by birds in mating. Some species—mainly those living on the ground—execute distinct, dance-like steps

while courting. Sometimes they are performed rather extemporaneously and enthusiastically, but at times these courting steps are repeated over and over in a painstakingly precise routine. Frequently, the male alone will perform this dance ritual in an attempt to ensnare the object of his affections. Often he will join several other members of his sex in a kind of "tournament dance" of martial character, during which the males pay almost no attention to the females standing off to one side—though the whole performance is meant exclusively for them. Among still other species, the females will take an active part in the dance, and frequently the partners will synchronize their steps and movements precisely to achieve almost perfect harmony, increasing their excitement to an ecstatic pitch.

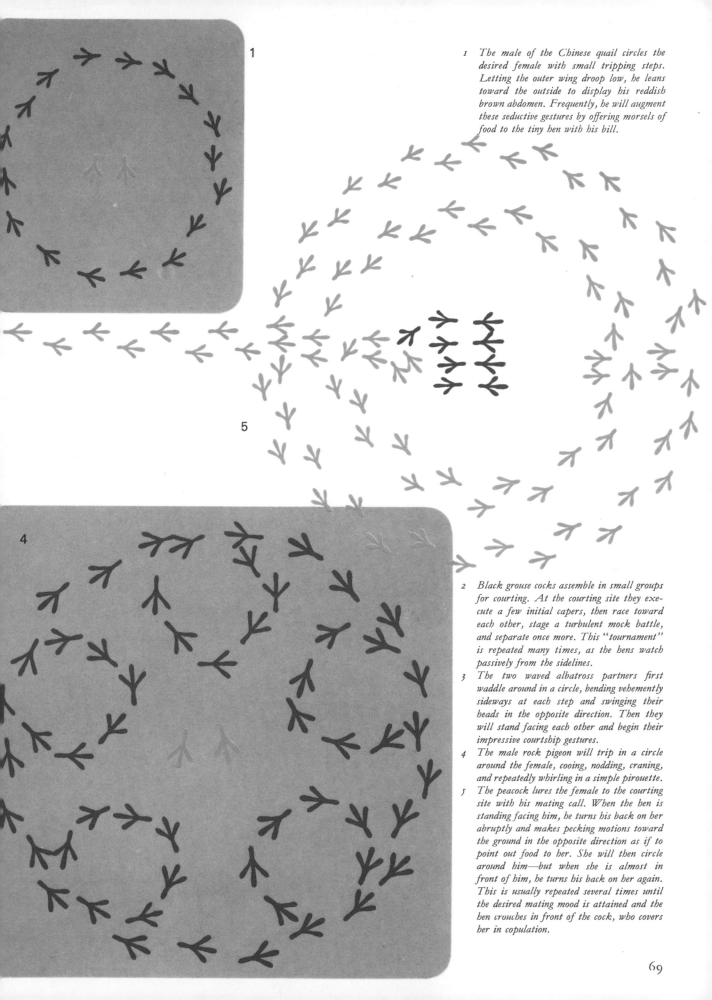

1   *The male of the Chinese quail circles the desired female with small tripping steps. Letting the outer wing droop low, he leans toward the outside to display his reddish brown abdomen. Frequently, he will augment these seductive gestures by offering morsels of food to the tiny hen with his bill.*

2   *Black grouse cocks assemble in small groups for courting. At the courting site they execute a few initial capers, then race toward each other, stage a turbulent mock battle, and separate once more. This "tournament" is repeated many times, as the hens watch passively from the sidelines.*

3   *The two waved albatross partners first waddle around in a circle, bending vehemently sideways at each step and swinging their heads in the opposite direction. Then they will stand facing each other and begin their impressive courtship gestures.*

4   *The male rock pigeon will trip in a circle around the female, cooing, nodding, craning, and repeatedly whirling in a simple pirouette.*

5   *The peacock lures the female to the courting site with his mating call. When the hen is standing facing him, he turns his back on her abruptly and makes pecking motions toward the ground in the opposite direction as if to point out food to her. She will then circle around him—but when she is almost in front of him, he turns his back on her again. This is usually repeated several times until the desired mating mood is attained and the hen crouches in front of the cock, who covers her in copulation.*

Mating Ceremony
of the Albatross

As a prelude to their mating ceremony, male and female waved albatrosses circle each other with slow, waddling steps, vigorously inclining their bodies to one side at each stride and counter-balancing themselves by swaying their heads to the opposite side.

After this introductory dance, the partners stand facing each other, and, with outstretched necks, they engage in a mock duel of beaks in which the beak-tips beat against each other in a drum-roll of short strokes.

The drum-roll completed, they often nibble at the points of their respective beaks. Suddenly, one of them will jerk its neck up and open its beak wide. The other bird will sometimes continue what it is doing, but usually ...

... it will immediately snap its head up, too, and open its beak. Both birds will hold this position for a moment and then will simultaneously flip their beaks shut with a resounding clack.

Wooing among albatrosses is particularly impressive despite—or perhaps because of—a lack of splendor in their plumage. The fascination of their display derives from their grotesque dances, made up of odd movements and often performed in rhythmical and harmonious unison.

Except for one species, which nests close to the equator, the waved albatross, albatrosses breed in the southern hemisphere, mostly near the Antarctic Circle. The is necessitated by the size of these birds, which approximates that of a turkey. The largest member of this family, the wandering albatross, has the longest wingspan—ten feet or more—of any bird.

mating games of the waved albatross have been observed nowhere in the world except on one of the Galapagos Islands in the Pacific Ocean. After more than two months' brooding on their single egg and nearly seven more months devoted to the upbringing of their offspring, the parents will leave the island, dissolve the marriage, and undertake vast migratory flights over the oceans, returning only every second year to entice prospective partners with their odd contortions. The long period of raising the young

This game is repeated at irregular intervals, with occasional variations: a soft chattering of beaks with lowered heads, a symbolic preening of the shoulder plumage, etc. Toward the end of the ceremony, both will point their beaks at the ground, as if to indicate the nesting site.

Finally, they will settle on the ground and caress. A gentle scratching of each other's neck feathers is what they seem to like best of all.

71

STRANGE COURTING CUSTOMS

◀ *The male golden-backed weaver has ingeniously braided a basic nest structure in a papyrus swamp. When a female approaches, he will suspend himself, head down, from the nest in the courting pose shown here, while sounding piercing calls and savagely beating his wings. The female, lured in this way, will inspect the nest, slipping inside to test it—and, in most cases, copulation will take place forthwith.*

GREAT CRESTED GREBES

The great crested grebe indicates its readiness for the courting dance by crouching low, raising its crest, spreading its wings, and inclining forward.

In the "cat pose," and with calls of "gah-gay-gay..." gradually increasing in volume, the partners approach each other.

Then they will suddenly rise high out of the water, assuming the vertical "penguin pose."

The water ballet of the great crested grebes, during which they select their partners in the early part of the year, is a magnificent spectacle. The tints of the plumage, which range from grayish to reddish brown, are especially vivid at this time, and the feathered crest is fully formed. Both sexes have the same colors, and since they also show identical courting behavior, it is almost impossible· to distinguish the male from the female. Only the actual copulation would reveal this information, but that takes place in the seclusion of the nest. The wonderful amorous prelude, however, is publicly displayed on the water. The assumption that copulation occurs at the climax of the courtship ritual, and that the

The dances of the cranes are among the most splendid of all bird performances. These expressive ceremonies can be observed throughout the entire year, and the larger young birds participate in them even before they have reached sexual maturity. This has led to the erroneous conclusion that these dances are merely an expression of the joy of living. Actually, they are an integral part of the courting ritual, performed exclusively for the purpose of mating and holding the couples together. The participation of the youngsters during puberty is probably nothing more than an exercise in the development of adult behavior patterns. Long before reaching sexual maturity, the young of many birds per-

This breast-to-breast position is the climax of the courting ritual, in which the partners frequently offer each other nesting material.

The courtship ends abruptly, the noise dies away, and the newly paired birds, in normal swimming position, turn their heads with a jerk in opposite directions!

CRANES

Cranes repeatedly bow to each other during their very beautiful mating dance. They run around one another intermittently, each "craning" toward the other.

These gestures are interspersed with capers which give the appearance of flightless birds attempting to take wing.

"penguin pose"—breast to breast—is the mating position, has proven to be erroneous. The offering of nesting material, seems to be of special importance. The couple submerges shortly before the final phase of nearly every one of the endlessly repeated courting gestures, only to reappear forthwith for the final scene, carrying bits of vegetation in their bills.

form actions which are clearly part of the reproductive behavior, practiced in a playful manner. At the age of only a few months, for example, young cattle egrets will display courting gestures and attempt to build nests. The calls of the crane couples—the "triumphal cries" uttered alternately by the two partners—also are practiced by adolescent cranes.

The mating patterns of the white storks are extremely complicated. They leave their African winter quarters as early as January and start to migrate back to Europe. It will take them almost two months to reach their breeding site, however, because their flying technique allows them to progress only very slowly. While most other migratory birds beat their wings indefatigably, as if they could hardly wait to see their home grounds again, the storks are complacent gliders. Circling in the currents of updrafts, they will allow themselves to be raised to a certain altitude on motionless, spread wings like birds of prey; then they will glide for some distance in the desired direction and look for the next updraft. When these sociable wanderers approach their breeding districts, they will gradually dissolve their groups, and each individual will head for its own aerie. The males seem to be more impatient

and in a greater rush, since they customarily take possession of the nest before the females arrive. As a rule, the "lady of the house" flies in a few days later, and the partnership, interrupted in late summer, can now be resumed and cultivated. Of course, it frequently happens that the homecoming reunions meet with some initial difficulties, primarily from young storks who are still unattached, have no home of their own as yet, and attempt to take possession of the strange aeries for themselves. If the intruder is a young male, the older bird will rout him energetically, and emphatically, and the matter is usually settled rather quickly. But if a young female arrives, the master of the aerie evidently has no objections; he will greet her enthusiastically and appear highly pleased with her presence—provided his partner of the preceding year has not arrived yet! In that case, however, he will never even get to this point: in a rage, the indignant mate will rush at the younger female and will, if necessary, also make unrestrained use of her dangerously pointed beak. But in their youthful obstinacy, the intruders often return to the aerie again and again, and violent, bloody, and sometimes mortal battles may ensue.

If the couple is not disturbed, the partners will begin by preparing the nest for the eggs. They will poke and tuck to loosen the old padding and will collect new nesting material, mainly to be used for upholstering. This joint repair work enhances contact and plays a part as introduction to mating, since the courtship is composed of certain formalized nest-construction gestures.

The activity during the construction is repeatedly interrupted by these ritual gestures: nesting materials are shifted quite unproductively from one place to another, twigs are pushed back and forth, and there is a great deal of poking about for no particular purpose except to make the characteristic motions which stimulate the mating urge of these birds. The drawings on these pages show how complicated the preliminary ceremonies for the actual mating really are.

Of course, the number and variety of the gestures may alter considerably the appearance of this mating game.

Larger bird species apparently have greater difficulties in copulating than smaller ones. Their introductory gestures are more complicated and much lengthier. As a rule, it takes far more time until the first copulation occurs, and afterwards the intervals between unions—which, incidentally, are generally quite numerous—are greater. Thus the connubial posturing of the stork is no isolated case. Equally impressive rituals are encountered among many other large birds as well.

The ostrich, for example, courts in two stages. A preliminary courtship takes place at the beginning of the breeding season, in order to isolate the future family group from the flock. For this purpose, the huge cock walks away from the community, alternatingly beating his wings to lure the desired hens. If he does not get

THE RING-NECKED PHEASANT

The courting ring-necked pheasant stands alongside the hen, spreading the wing on her side down to the ground to impress her with its colorful iridescent splendor.

bottom of the nest, and by striding with dignity around his mate. When the female has risen, they will circle each other, intermittently—and often simultaneously—pointing their beaks at the bottom of the aerie, and scratching each other's necks and wing roots. Finally, they will copulate with spread wings, and then each will scratch and preen itself, in what almost looks like a gesture of embarrassment.

results with this method, he will herd his chosen partners away from the group one at a time. Then he will march with the small conquered harem to his breeding area, where the main courtship takes place. The lengthy, complicated mating process is repeated several times with each hen. It is a fascinating spectacle for an observer. As soon as the male gets into the mating mood, he will herd one of the females away from the rest and begin to peck food with her. Gradually the two will adjust the rhythm of their pecking motions to each other, and simultaneously the action becomes ritualistic: from now on, they will only make a show of taking the food. These synchronized symbolic motions evidently increase the sexual excitement. Next, the male will begin to beat his wings alternately and to poke around in a nesting site in a gesture which is also ritualistic. He will drop to the ground

and throw sand to one side as if to dig a nesting trough, twisting his neck in all directions and uttering low rumbling calls. Then he will suddenly jump up again and at the same moment the hen will drop to the ground to let the greatly excited cock copulate with her.

In addition to the acoustical courtship rituals of the feathered vocalists—large and small— along with the expressionistic dances and the vain display of magnificent feathers (all of which are performed on the ground, among branches, or on rock projections), many species also use the air as the courting site of their choice. The diurnal birds of prey, in particular, will demonstrate their affections in elaborate aerial performances of ravishing beauty. Despite

the fact that in some of these species the couples remain together for many years, they perform the same courtship rituals annually, expressing their ardor by flying loops and spirals, wild chases, and whirling dives. Some birds which normally tend toward taciturnity become extremely vocal in spring. During the mating season, some types of buzzards will "meow" at each other for days on end. Other varieties will sound whining, croaking, or barking calls. Some merely manage peeping sounds which seem entirely out of character among these proud birds, and others remain completely silent during courtship.

THE COMMON CURLEW

The common curlew lures the lady of his choice with raised wings. As soon as he has succeeded, he symbolically shows her the nesting site in his area by pointing downward with his beak. Then he presses his breast into the sand and turns in a circle, thus ceremoniously making a depression in the ground for the speckled eggs. In this way, he demonstrates his willingness to mate and to stay with his partner for a long time to come.

But it is not only the royalty of the air which takes to courtship on the wing. A great many ground birds prove to be amazing aerial acrobats at mating time. Representatives of the fowl species, such as the male ptarmigan, will introduce short, steep flights, accompanied by loud calls, in their ground ritual. And numerous waders, earthbound at other times, display some of the wildest, most extravagant aerial rituals of all birds during courtship. Woodcocks will fly wild zigzag courses, particularly at twilight; curlews, after a steep ascent, will glide slowly to the ground, trembling wings slanted upward, and then continue to court with expressive gestures. And anyone who has witnessed the courtship flights of the black-and-white lapwings, with their caprioles, flourishes and whirls, loops and spirals, inverted flights and fluttering dives, will find it difficult to avoid the impression that these courting couples are exceedingly happy to be alive.

The squacco heron, only as large as a pigeon, makes the most of his small body at mating time. To attract a female, he poses impressively, making a beautiful display of the ornamental feathers on his head and neck.

The Eastern Ruffed Grouse

The Hummingbird

If a female hummingbird is sufficiently impressed by the demonstration flights of a male, she will join him in a courtship flight. Together the couple will make looping, curving flights, in the course of which the pair will repeatedly stop suddenly in mid-air and hover for a few moments, closely facing each other.

Gray Herons

When the male gray herons return to their breeding areas at the start of the mating season, they immediately occupy the old nesting sites, usually located in colonies in treetops or occasionally in reed banks. The first arrivals usually select the larger nests; male herons arriving later have to be satisfied with the smaller ones or build new ones if necessary. When a female heron is in sight, every owner of an aerie will immediately muster all his ability to lure her his way, drawing himself up in "pole position" or inclining his body forward, bending his neck back and pointing his beak up. And intermittently, he will sound his coarse mating call. When a female heron eventually lands at the nesting site, he will begin immediately to tinker about the aerie. Actually, he does not accomplish much, but in order to woo his mate he must first of all demonstrate symbolically that he is willing to be a good father. Evidently, these ritualistic gestures convince the female of his good intentions, because she will soon participate in this symbolic nest construction.

The Lapwing

In his area-occupation flights and aerial courtship, the male lapwing performs the most daringly reckless aerobatics! He will climb steeply, flip over again and again, and plummet down, turning spirally about his own axis, whirling and fluttering.

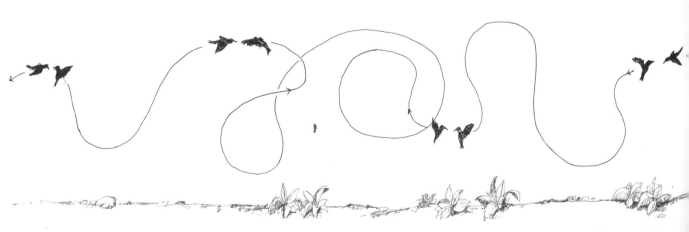

All year long, the male vermilion flycatcher, living in the jungles of South America, keeps his breeding area to himself and his mate by virtue of his song. Most songbirds stake their claims in the same manner, but as a rule they do so only during the breeding season. The vermilion flycatcher shown here is still a young fellow and his plumage is not fully colored as yet. Later on, his breast, his abdomen and the upper part of his head will be a brilliant ruby red.

Be its plumage ever so beautiful
and its flight ever so marvelous,
it is the song
of the bird,
above all,
that gladdens
the hearts of men.

Among more than half the manifold bird species, it is song which plays a dominant role in the mating game.

For a great many birds, song is of the utmost importance when they lay claim to their breeding areas, and, therefore, of the greatest significance for the entire process of propagation. Songbirds must disperse all over the countryside at the beginning of the mating season, since each couple requires a territory of its own, and one with an adequate food supply for itself and its young: insects and their larvae, worms, spiders, and much more besides. And this territory must be held if the offspring are not to starve to death in the nest. The size of the breeding area varies quite considerably, of course, not only according to the species, but also depending upon the locality and the conditions obtaining in a particular year.

As soon as the male songbird, in search of a mate, has found an area to his liking, he stakes his claim to it loudly and volubly. Initially, this song—which he will pour forth indefatigably, from early morning until dusk— is nothing more than an acoustical fence intended to keep out all male members of his species. But soon it will also double as a lure for a female, which, presumably, will look upon the racket as music to mate by.

The song—its melody, rhythm, volume, and pitch—varies according to the species, since the males it is intended to keep out and the females it is supposed to entice are all representatives of the same species (several different species can breed in the same area without difficulty and without contending with one another for food; some will seek it among the treetops, others on the leaves of bushes, in the grass, in the foliage, or in the air).

So effective is this song, however redundant it may be, that only in the very rare cases will another male in search of a breeding area attempt to invade occupied terrain. Even when a male is placed in a cage in his own area, he will continue to hold the territory with his voice alone, but when the same cage is moved to another male's breeding ground, its occupant will crouch in fright in a corner and will stop singing. In an experiment, one such bird even died of a heart attack when the rightful resident of the area assaulted the cage.

Owing to the vigor of the song, fights almost invariably can be avoided in occupied breeding areas. But other species, which are not at all "musical," also show resourcefulness. They will arrange things among themselves by means of ritualized war dances at the borders of their territories. But if occasionally violence does occur—as is the case among coots, for example—peace is generally restored forthwith; after a short energetic scuffle the intruder will make it clear that he considers himself to be the loser and give in. This surrender may be accompanied by gestures of appeasement and humility which are characteristic of the species, or simply by flight. Only on rare occasions is this a sign of actual inferiority, however, because the situation immediately is reversed if the "victor" of one moment should cross the territorial boundary during the rout, thus intruding on the area of his "defeated" adversary. A bird will nearly always feel less secure in another's territory than in his own, and therefore will generally acknowledge defeat rapidly when away from his home ground.

Pages 80/81

Comparative sound spectrography of songs and calls reveals the greater structural simplicity of the calls of the blackbird, the short-toed tree creeper, and the tree creeper. A single bird species may have more than twenty distinct calls, each with a different meaning. With some species, communication actually begins within the egg, as in the case of Bobwhite chicks, which use clicking noises in preparation for simultaneous hatching. Guillemot chicks inside the egg learn to recognize the individual voices of their parents.

The warning calls of songbirds facing enemies such as owls, diurnal birds of prey or beasts of prey are noisy and effective— when songbirds discover a tawny owl sleeping in daytime, their calls of alarm raise an infernal racket. The begging calls of youngsters induce their parents to bring them food. Birds migrating at night maintain contact with each other by means of calls resounding through the nocturnal skies. There are other contact calls which serve to keep couples or flocks together, and many more calls which facilitate — in some cases even initiate—certain forms of community bird life.

I MAIN SONG OF MALE

BLACKBIRD

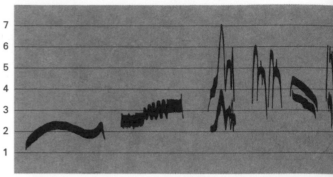

This double page shows the sound spectrograms of the songs and calls of different bird species. These vocal utterings vary from one variety to the next, and a considerable difference can be clearly noticed even among such closely related species as short-toed tree creepers and tree creepers. The numerical scale at the head of the tables indicates the pitch in kilocycles (1 kc = 1000 cycles per second).

SHORT-TOED
TREE CREEPER

I   The first section represents the main song of the male when occupying his area at breeding time; the song may vary considerably within any given species. Woodpeckers use special drumming signals instead of song.

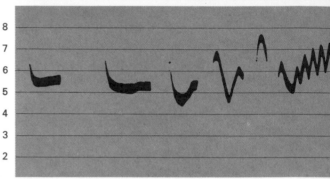

TREE CREEPER

NUTHATCH

II   The alarm calls shown here are used to warn of impending danger. Other bird species—and even mammals—often will understand and heed these alarm calls.

III   Begging calls are used by youngsters to demand food from their parents. The begging calls of most late nestlings are very low in volume, particularly during the first few days, which keeps the young birds from being readily discovered by predators.

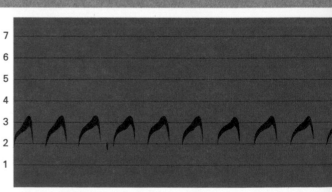

GREAT SPOTTED
WOODPECKER

IV   Contact calls are an important means of keeping partners in touch with each other, since forest, bush, and reed dwellers continually lose sight of each other. Similar calls help to maintain flocks intact.

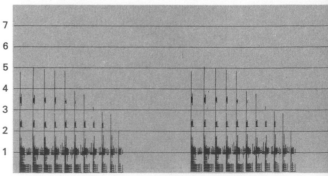

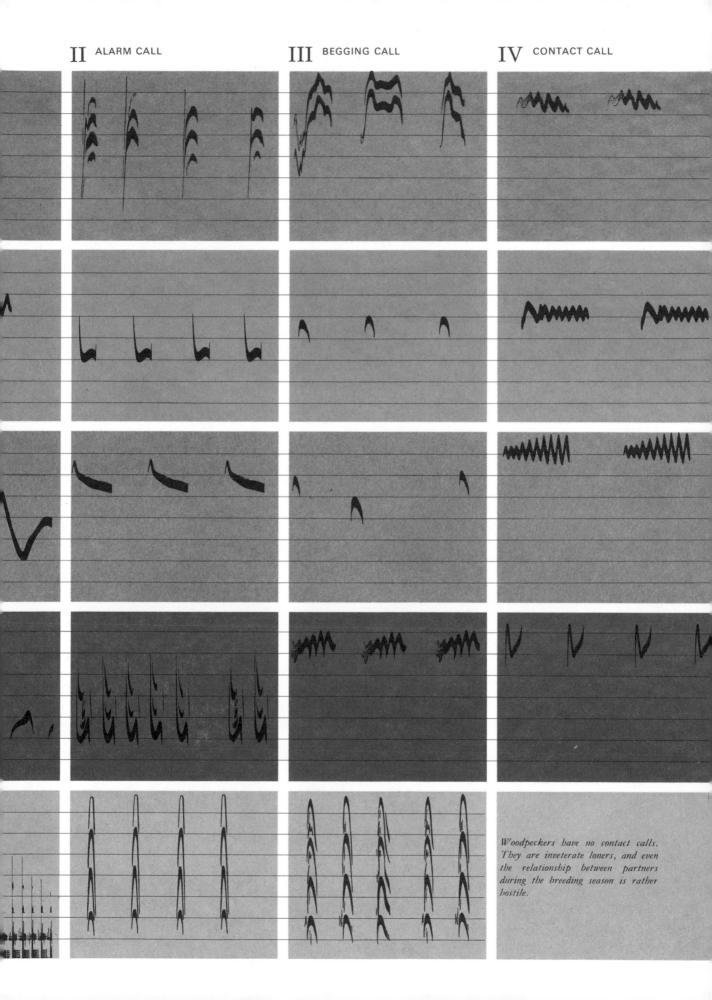

II ALARM CALL   III BEGGING CALL   IV CONTACT CALL

Woodpeckers have no contact calls. They are inveterate loners, and even the relationship between partners during the breeding season is rather hostile.

THE BIRD
OF GODS AND KINGS

The most famous figure among courting birds is, without a doubt, the peacock displaying his spread fan. The wedding costume of this bird has gained him the attention and protection of man throughout the world.

He is the oldest known ornamental bird. As early as two thousand years before the birth of Christ he was taken from his Indian habitat into the Persian area and thence into the Mediterranean. In Greece and Persia particularly, he soon played as important a role in man's rituals as he did in India.

It was not the splendor of his plumage which was deified, but the eyelike markings of the spread peacock fan. Man believed in the magic power of the eye as a defensive and protective force, and the eye symbol was regarded as the most potent charm in averting misfortune and danger.

In India the peacock is the symbol of Krishna, who sees all and must have countless eyes. Representations of this deity very often take the form of the peacock; consequently, the wild peacock is never killed in his native land. Because of this protection he often loses his shyness, living fearlessly in the midst of human habitations—an old tree in the middle of a village will often furnish the peacock with sleeping quarters. Some of the characteristics of this bird tend to confirm the Judeans' belief that he is a sacred protector of man. He utters loud warning cries, for example, when he discovers a roving tiger; he kills and

The peacock's little crown may also have contributed to his widespread acceptance as a royal bird. Generations of kings in many lands have kept this bird at their courts, and even Alexander the Great had peacocks in his garden.

devours every small snake he finds, and it is said that there are hardly any of the dreaded cobras in areas inhabited by peacocks.

The throne of the Persian Shah is called the Peacock Throne, and the coronation mantle of the Empress Farah is decorated with lavish embroidery representing the eyes of a peacock. A sixteenth-century Persian illustration shows Mohammed riding through the seven spheres of heaven on his mare Burak, a deity with the body of a horse, the crowned head of a human being, and the tail of a peacock.

In Greek mythology the peacock is consecrated to Hera, the Queen of the Gods. It was she, who, after the death of Argus, the hundred-eyed giant, placed his eyes on the feathers of the peacock.

Although we know today that the peacock is a thoroughly earthly creature and that the eye markings of his magnificent fan merely serve to attract his partner and put her in the mood for mating, this bird remains a creature of mystery and a delight to the eye of man, in his unquenchable thirst for beauty.

After the rain
the blue peacock sounds his shrill
mating call.

Plain and earth-colored,
his harem emerges
from the bush into the clearing.
Excitedly he stretches his iridescent throat,
and rises high his crown.

The Moslems regard the peacock as the symbol for the dualism of the male soul because of the difference in his appearance at rest and in the courting posture. A legend, also of Moslem origin, has it that the peacock was evicted from paradise together with Adam and Eve and the serpent, that he lost his delightful voice at that time, and that ever since he has been able to utter only ear-splitting shrieks.

Christians also have attributed symbolic significance to the peacock. The shedding and subsequent renewal of his magnificent ornamental feathers have been regarded as a symbol of resurrection and immortality.

To this day, peacock feathers are worn in many places as a protection against the "evil eye." Stylized representations of the eye are often found, again as defensive symbols, on ancient military armor, particularly near such sensitive areas of the body as the knees and the shoulders. And many national emblems—those of Nepal, Burma, and North Korea, for instance—include representations of peacocks.

Eye markings occur quite frequently in nature, and almost invariably they have a deterrent, defensive function. They are found on the wings of butterflies, on caterpillars, beetles, fish, reptiles, and many other animals. The names of these animals are often derived from the most famous wearer of mock eyes: there are diurnal and nocturnal peacock butterflies, peacock fish, and so forth. Many birds are also named after this "divine animal"—the peacock pheasant and the pea dove are but two examples.

The Courtship Ritual

In his native habitat, the peacock breeds during the rainy season, when he becomes particularly vociferous and serenades his mate with piercing cries.

The peacocks, like most other birds, behave unsociably toward each other at mating time. Each male looks for a breeding area and tries to establish a harem of from two to five hens. Rivals seeking to elope with one or another of the hens are unceremoniously routed. The bonds within the matrimonial group, however, are rather slack. Each of its members goes more or less independently in search of food. The modestly colored females enter the male's courting area whenever they please. As soon as the cock catches sight of one of the hens, he begins his incomparable courtship ritual. Facing the hen, he erects his tail feathers in a dazzling display of ornamental plumage.

Now, as the hen approaches, the male's excitement mounts. He unfolds his feathers to their full splendor, until they face forward in the shape of a paraboloid, lowers his wings with shoveling and jerking motions until they almost touch the ground, and shivers repeatedly, causing all his plumage to rustle. But when at last the hen stands facing him, a strange thing occurs: while retaining his ornate dignity, the cock turns his posterior toward the hen, who runs around him so that they are face to face.

On the following pages: *The pigments of the peacock's plumage are mainly brown. His colorful splendor, as is the case with other iridescent animal displays—butterfly wings, and the wing cases of beetles, for example—is really an optical illusion caused by light refraction in the surface structure of the feathers which changes according to their angle. The red lower-frequency components of the light are absorbed, while the high-frequency blue radiations are reflected by countless tiny air bubbles.*

The cock makes pecking motions toward the ground in front of the hen, thus symbolically pointing out food to her. In this way she is constantly enticed toward the center of the paraboloid. Eventually she crouches in this spot, whereupon the cock suddenly whirls and drapes himself completely over her and initiates copulation.

The fan display of the peacock consists of greatly elongated dorsal feathers which he carries behind him like a train when they are in their normal position. Each of these feathers has a conspicuous "eye" near the outer end, and since the feathers vary in length, the eye pattern is distributed over the entire surface of the display.

The twenty light brown tail feathers which support the fan display are much shorter. Peahens frequently present such a display too, especially the younger birds. Even very small chicks attempt to do so quite frequently; they shrug their small wings in a grownup manner and form a tiny fan with their short tail feathers. The fan display, therefore, is not a mere courting posture reserved for the mating cock. Moreover, the cock will assume this same impressive posture outside of the mating season—although less frequently, and with less emphasis than during courtship.

BIRDS AND THEIR EGGS

The diversity of birds is unbelievable, but they do have one thing in common: they all lay eggs—the minute hummingbird and the giant ostrich, the "amphibious" penguin and that majestic denizen of the mountains, the eagle.

Among fish and amphibians there are species which give birth to living young. The offspring of some lizards and of some New World boas develop within the mothers' bodies, and the viper's very name is derived from the word "viviparous," which means giving birth to living progeny. On the other hand, among the mammals—classically viviparous animals—there are two oviparous species. Both the duck-billed platypus and the echidna lay eggs—a discovery not made until 1884, in Australia, where unusual animals abound.

But *all* birds lay eggs—although the eggs of no two species are exactly alike. They vary in size, color, and markings, in shape and surface structure. In these qualities and in everything else concerning their eggs, the birds demonstrate once again their almost limitless variety.

A complete brood may consist of a single egg or as many as eigh-

Their delicate tints,
their immobility,
are reminiscent
of agate,
marble, and jasper
from the bowels of the earth.

But what appears
as lifeless mineral
is but the shell
whose dark confines
hold
the greatest of all miracles:
the promise of life.

teen eggs. Many birds deposit one egg every morning, some lay in the afternoon, still others lay at intervals of from two to five days. Some birds even pause more than two weeks between eggs. In some species only the female will brood, in a few only the male takes care of incubation, while in many species the partners will alternate. Occasionally, a bird will sit on the broods of one or two neighbors of the same species along with her own eggs. Some species smuggle their eggs singly into alien nests. And there is one group of small birds resembling chickens which follows a particularly old brooding procedure: they bury their eggs in the earth like their quasi-reptilian ancestors. This method is, however, far more laborious than the usual process of incubation by body heat, for it involves a scrupulous surveillance and regulation of the temperature in the underground incubating chamber. Later we shall become better acquainted with the so-called "mallee fowl."

Hornbills have found an excellent method of protecting their eggs: the parents will wall up the entrance to a sufficiently large cavity in a hollow tree with clay, leaving only a small opening through which the brooding female can be fed by her mate. After a short time, the mixture of clay and saliva will be as hard as rock.

EGGS ON THE RAFT

The great crested grebe woos his mate in the charming ritual of a water ballet. Once the relationship is established, the birds gather together a pile of plant fragments, build a floating home, and celebrate their nuptials on their raft-like nest. The female then lays from three to seven eggs. Initially, the eggs are almost white, "impregnated" with a chalky calciferous matter which protects them against the dampness. Soon, however, they take on a dark olive drab coloring from the rotting, water-soaked nesting material. As soon as the first or second egg is in the nest, the couple begins to brood alternately. When danger threatens, the brooding bird rises, quickly and skillfully covers the eggs with a beakful of nesting material, and then drives away noiselessly and secretively. In this way the eggs are protected from cold, heat, and the predatory eye of would-be nest robbers. The parents wait within sight of the nest until the coast is clear and then return cautiously to the raft.

On the following pages: Eggs of various birds in the corresponding nests. The measurements are axial and obviously may vary considerably within any given species.

1	ARCTIC TERN	1.57 × 1.14 in.
2	SONG THRUSH	1.00 × 0.73 in.
3	MAGPIE	1.38 × 0.95 in.
4	CHAFFINCH	0.78 × 0.55 in.
5	CORMORANT	2.48 × 1.57 in.
6	YELLOWHAMMER	0.80 × 0.57 in.
7	STARLING	1.17 × 0.83 in.
8	SNOW BUNTING	0.78 × 0.55 in.
9	COMMON GALLINULE	1.60 × 1.14 in.
10	GOLDEN ORIOLE	1.17 × 0.83 in.
11	GARDEN WARBLER	0.80 × 0.52 in.
12	LITTLE GREBE	1.49 × 1.02 in.
13	LESSER WHITETHROAT	0.61 × 0.41 in.
14	GREAT GRAY SHRIKE	1.03 × 0.76 in.
15	LESSER GRAY SHRIKE	0.98 × 0.68 in.
16	OXEYE TIT	0.72 × 0.53 in.
17	TREE SPARROW	0.76 × 0.56 in.
18	ROBIN	0.76 × 0.58 in.
19	BLACKBIRD	1.10 × 0.76 in.
20	PARROT CROSSBILL	0.91 × 0.63 in.
21	LONG-EARED OWL	1.57 × 1.26 in.

Returning to the nest, the great crested grebe carefully removes the leaves covering the eggs.

The brooding bird, always suspicious, is continually on guard.

While other aquatic birds, such as geese and ducks, almost invariably consummate their partnership in the water, the great crested grebe mates unobtrusively and secretly on the nesting raft, the female lying stretched out flat under the male, poised erectly above her.

As soon as two thirds of the chicks are hatched, they leave the nest with their parents. The remaining eggs are abandoned and the unhatched young, although almost ready to emerge, are left to perish.

22	GREENFINCH	0.85 × 0.63 in.
23	MALLARD	2.24 × 1.57 in.
24	KITTIWAKE	2.16 × 1.57 in.

4

5

6

10

11

12

16   22 ▼

17   23 ▼

18   24 ▼

The brooding swan mother rises from time to time; and just when one might think she is about to leave the nest, possibly in search of food, she does something quite extraordinary: she slowly slips her beak under the eggs and turns them carefully! She will devote herself to this task for one or two minutes and then snuggle down on the nest once more. If the eggs are not turned periodically, the settling yolk might stick to the lower portion of the shell and thus be damaged. The brooding bird does not know this, of course. But does this not make it all the more amazing that she will skillfully provide for the necessary rotation of the eggs at regular intervals?

The mute swan is one of the rare birds who does not produce her numerous eggs at twenty-four-hour intervals. Her six to eight eggs are laid from two to three days apart, so that it will take from two to three weeks until they have all been deposited. During this time the bird will not sit, for brooding must not begin until the last egg is in the nest.

How does the swan know when the time has come to brood? Those mysterious messengers called hormones, which circulate in the bloodstream, will tell her when, and for how long, she must sit on her eggs.

During the long egg-laying period, the swan parents, who live in exemplary monogamy, remain constantly in the immediate vicinity of the nest. As soon as the last egg has been laid the female receives the "command" to brood. As is customary in the duck family, the female bears sole responsibility for warming the eggs. Only among black swans, tree ducks, and Australian magpie geese does the male share in this task. Still, the male mute swan is most conscientious. He is always in close attendance to his

brooding mate, and she can doze confidently on her nest. The male swan will defy death in defending the nest against all approaching danger. There is hardly another bird whose defensive behavior is bolder or more ferocious in appearance. He will lunge at man or

beast, making angry noises and beating his wings; he has even been known to put large dogs to flight.

That the swan will stage such a death-defying defense of his brood might logically lead to the conclusion that he knows his own eggs. This, however, is not the case. When an egg lies on the edge of the nest, he will roll it back towards the others. But if it lies a few feet away, he will no longer bother with it.

The husband is exceedingly solicitous, however, that the nest—actually a large floating raft—remains intact. He continually bites off dry reed stalks in the vicinity and places them next to the nest for his mate. Without rising she picks up the stalks with her beak and builds them into the nest. In the course of the incubation period the reed raft will grow into a considerable pile which may attain a diameter of six and a half feet and a thickness of over three feet. As a result of this continuous inforcement the eggs normally remain dry at all times, even though the nest gradually sinks lower into the water.

The female mute swan dozes thirty-five days on her eggs, half anesthetized by a strong dosage of tranquilizing hormones. Since she does not start brooding until the last egg has been laid, the development of the embryos, being contingent upon warm temperatures, begins simultaneously in all of the eggs. Consequently, the young will be ready to break through the protective shell all at the same time.

There lies the egg, inanimate and alien,
beside its lively parent.

What tremendous power
can constrain a bird,
yearning for flight,
to brood impassively
on its eggs?

ity. From dawn to dusk they are constantly in motion, except for a few short naps and a longer rest at noon. Their temperament corresponds entirely to their rapid pulse and to a body temperature of almost 106.7° F. But when all the eggs have been laid, there is an abrupt change in this restless life: the bird must brood. It must remain motionless during many hours of the day; temptations, irresistible at other times, must be resisted. Passing insects, deli-

furnish the nesting site with a few small shells or stones or bits of vegetation.

After the first egg has been deposited—provided the brood consists of more than one egg— most expectant parents will continue their inexhaustible activ-

The razor-billed auk, who breeds on a narrow, bare rock shelf, cannot hatch a large brood: it must content itself with a single egg.

The first matings occur as a rule during the last phases of nest construction, and the first egg is usually deposited as soon as the nest is completed. Naturally, this only applies to those birds who "marry" their partners, and even among them, only to those who build proper nests. The razor-billed auk of the northern oceans builds no nest at all—the female's one and only egg is laid on bare, unprepared rock. Most other birds, even the hastiest among them, will at least scrape a shallow depression into the ground or

Half a dozen reed warbler eggs lie well protected among the reeds in this expertly constructed pile dwelling.

Top right: While its relatives breed on the ground, or, at best, pile up a few seaweeds to make a shapeless nest, the red-footed gannet tries its luck at building a twig nest in a bush. These structures often are so insubstantial, however, that the light-colored eggs can be seen shimmering through the bottom of the nest from below.

cacies which at other times are voraciously devoured, must be ignored. No longer may every alarming sound trigger headlong flight. In short, actions which have become second nature must be suppressed. This sudden readjustment from maximum activity to almost total inactivity is without doubt one of the greatest accomplishments of the bird within the eventful course of the year.

How is it achieved?

Is it parental love?

Surely not. The brooding bird is not aware that its offspring will hatch from the hard ovoid objects on which it sits so patiently.

Of Eggs and Breeding

The bird is an animal of great sensitivity and very little intelligence. These qualities are entirely in keeping with the nature of this restless creature. When a bird in headlong flight is suddenly confronted with either danger or food, there is no time for deliberation; it must do the right thing instantly. With lightning speed it must dodge its enemy or swoop down on its prey. Fractions of a second may mean the difference between life and death, nourishment and hunger. It is instinct, too, that renders it capable of undertaking long migrations at the proper time.

In the course of their evolution, birds have increased their sensitivity to both external and internal stimuli, as well as their capacity for appropriate reaction. This is not to say, however, that birds are incapable of learning. Some species, on the contrary—particularly parrots and crows—are extremely docile. Moreover, the greater the number of a bird's acts, the more intelligent he must be in order to coordinate them properly.

Incubation is purely an instinctive act. As in most bird activities, this instinct is triggered by a special stimulus. As is the case with a great many stimuli, it is present only during certain seasons. The sight of the eggs alone generally will not suffice to generate the propensity for incubation. The bird must first be in the "mood" for brooding. This mood is aroused through an increase of certain hormones contained in the blood. Thus, not only the nerves but also the blood vessels serve as the communications system for the stimulus to which the bird reacts "automatically."

It is the hormones which compel the bird to remain sitting on its eggs until they are hatched.

In most cases it is not difficult to account for the various egg shapes. The kiwi embryo, for instance, requires a large egg, because the bird does not hatch until it has reached a very advanced stage of development. This unusually large egg, whose weight of one pound represents nearly one-seventh the weight of the bird itself, would hardly clear the oviduct and the depository orifice, the cloaca, if it did not have an elongated form. The expediency in the shape of auk eggs, pointed like a child's top, is even more obvious. As they are frequently deposited on narrow rock ledges, they might easily roll off whenever the brooding bird makes a hasty departure. But their conical shape causes them merely to roll in a small circle when they are set in motion.

For most of us, the egg is such a common sight that we hardly ever realize how marvelous it is. The embryo of the mammal is attached to the mother's circulatory system by means of the umbilical cord, and the red fluid provides it with everything needed for its development: nutritive substances, water, oxygen. The bird embryo, on the other hand, is cut off from the lifegiving bloodstream of the mother during its first stage of development and left to thrive in its isolated enclosure. All energies and body-building substances as well as the required fluids must be contained in the egg. But outside assistance is needed nevertheless. Warmth, more or less constant humidity, and oxygen must be provided. The sitting bird usually supplies the necessary warmth and humidity. Innumerable pores in the eggshell and an air chamber at the blunt end of the egg allow for the required oxygen exchange. Shortly before and during the laying of the eggs most incubating birds develop "brooding spots," formed by the loss of down feathers, and at times a certain number of contour feathers, on their abdomen. The exposed skin areas (frequently there are three of them) have increased blood circulation and warm up to almost feverish temperatures. When the bird sits on its eggs, it spreads the still remaining abdominal plumage and comes to rest on the eggs in such a way that the bare, warm skin areas nestle against the upper side of the eggs. In addition, the lining of the nest protects the eggs against the cold from below.

Many birds manage without this additional protection, on the bare ground or in a nest so loosely constructed that it has almost no heat-insulating effect. But the perfunctory nest builders almost invariably lay only a few eggs, frequently no more than one or two, which naturally can be warmed with the belly alone far more easily than would be the case with a greater number.

Where both sexes of a bird species have identical or very similar plumage, both parents will generally share the task of incubation. Male birds with conspicuously resplendent plumage would betray the location of the nest too readily and therefore do not participate in brooding. However, this can only be considered as a rule of thumb. There are numerous exceptions. For instance, both sexes of the swan are equally conspicuous, yet only the female broods. On the other hand, the far more colorful males of some finch species assist quite actively in incubation.

The brooding relief is almost invariably accompanied by certain ceremonies. These may be very brief and simple, consisting, perhaps, of no more than a touch of the beak of the relieving bird on the back of the sitting partner. But they may also be complicated, precisely repetitive actions, extending over several minutes, particularly among the larger bird species. The transfer of nesting material frequently plays an important part in these ceremonies, and for this reason many nests are continually enlarged during the incubation period. I have observed a pair of great white herons who followed this pattern without deviation: each time the relief partner approached, it brought a reed stalk which the brooding partner inserted into the nest prior to departure.

The brooding relief usually occurs at rather irregular intervals. With some species, one sex will brood at night and the other in the daytime. Almost invariably, however, the female spends more hours sitting on the eggs than the male. The male pigeon, for example, sits from early morning until early afternoon on the eggs, which are nearly always two in number. During the remaining time the female provides the needed warmth. They relieve each other with extreme punctuality. Because of their exact sense of timing, the sitting pigeon may leave the eggs as soon as she has completed her allotted turn without waiting for the arrival of her partner. He will appear within minutes to take over the brooding task. This may, however, have unfortunate consequences. If, for example, the female has an accident while the male is sitting, he will nevertheless leave the eggs punctually as is his wont and will return the following morning to spend his time sitting on the eggs. In the meantime the eggs will have cooled off and died, but he will repeat this senseless game until the brooding instinct wanes a few days after the normal incubation period.

With most species who brood in pairs, however, the sitting bird will wait until it is relieved. In case one partner has an accident, the survivor may quickly go in search of food and return immediately to resume brooding. In this way, even a widowed bird is often capable of raising its offspring on its own, even though it must perform almost twice the normal amount of work.

If a female bird loses eggs during the laying cycle, she usually is able to produce additional eggs, since brooding does not begin, as a rule, until the full number of eggs is lying in the nest. It follows from this that she does not keep a "running count" of the number of eggs she has deposited, but rather of those present in the nest. The pigeon is again the exception. Even if one of her two eggs is removed, she will continue to brood without laying again. But if both eggs are taken away, the entire mating game will start again from the beginning. Cooing and nodding, the male pigeon will prance around his mate frequently, within half an hour of the loss, to invite her to mate with him once more. This time, of course, she will agree more readily than on the first occasion, and two days later the first egg may already be deposited in the nest. But this game will not be repeated too often—three or four times at the most—and thus the domestic pigeon has always played a rather insignificant role as an egg producer.

Many species will continue laying without interruption if eggs are taken from their nests. A jackdaw, which normally lays about five eggs, has been known to produce sixteen of them, and a wryneck has even managed to lay sixty-two eggs. Many birds will continue laying until they die from exhaustion. On this basis there have been attempts to decimate sparrows, magpies, and other birds systematically.

The red jungle fowl of Asia is a particularly good layer without incurring damage. Consequently, it has been domesticated for a long time, and over five hundred years ago it was exported from India to China. It is the species from which the numerous varieties of domestic chickens all over the world were gradually bred. Their egg production rate has been continually improved by means of selective breeding, and today a hen may produce considerably more than three hundred eggs per year. Few ducks have a numerically comparable output, and even those that have require far more feed to achieve it. This is why ducks are rarely kept as egg producers these days.

There is still considerable uncertainty as to what extent birds are able to identify their own eggs. By and large, the nest and its environment seem to impress themselves on the bird's memory more readily than the eggs.

In the case of the swan we have already seen that it will ignore its eggs, once they are removed by even as much as one yard from the edge of the nest, and we also know that it will calmly continue to

The illustration on the page at right provide an idea of the incredible variations in the shapes of birds' eggs. Those shown here are life size.

WHITE HERON

CORMORANT

GOLDEN ORIOLE

COMMON SNIPE

LESSER WHITETHROAT

OXEYE TIT

KESTREL

KINGFISHER

ALPINE SWIFT

OSTRICH

KIWI

HUMMINGBIRD

GREAT HORNED OWL

GUILLEMOT

brood on bottles which have been substituted for its eggs. Heron eggs have been replaced occasionally by square wooden blocks and were accepted and sat on without opposition. Herring gulls will not notice that brightly colored artificial eggs have been smuggled into their nests, and a canary hen will brood indefatigably on gravel, as long as it is approximately the same size as her eggs, although there is no similarity in color.

But many birds do know the exact number of their eggs, as long as there are only few of them, and some others, who produce larger numbers, are fairly well aware of the overall amount. The sooty tern, which lays only one egg, will sit on no more than one. If a second egg is added, though it may be of the same species and nearly identical in appearance, the bird will most probably be able to identify its own egg and will push the other one aside. Thus it is all the more astonishing that it will sit on a light bulb of about the same size that has been substituted for the egg in the recess of the nest.

Songbirds will ignore their eggs when they are moved to the extreme edge of their nest, and some birds refuse to brood when a cuckoo egg lies in their nest, even though it is similar to their own in size and color and though the female cuckoo has taken the precaution of removing one of the host's eggs.

Camouflage is the most widely used defense against nest robbers. Gallinaceous birds, ducks, nightjars, and many others remain crouched over their eggs in complete immobility when danger threatens and can hardly be differentiated from their surroundings. Many birds which brood on the ground will not fly off until the intruder is almost upon them.

When brooding bitterns are disturbed, they freeze in a "pole position" with neck and beak raised vertically. In this position, the yellowish brown neck blends marvelously with the vertical reeds around the nest. Some birds are so clever at dup-

ing their enemies that even a bird expert might be deceived by them. They will stagger to and fro, they will drag one of their wings as though it were broken, they will lie down occasionally, fluttering helplessly, and will permit a human being to approach them to within far less than the usual escape range. Only after these maneuvers have drawn the enemy a considerable distance from the nest, will they suddenly dart off. This ruse looks so deliberate that one is tempted to ascribe it to the bird's intelligence, although it is performed instinctively.

Wherever an opportunity has existed for birds to reduce their weight load, it has been exploited in the course of their evolution. This is evident in the nature of the reproductive organs. Female birds, like all female vertebrates, have two ovaries, but, with very few exceptions, the right ovary is stunted and has lost its function. Even among the few birds which develop two mature ovaries, only the left one is equipped with an oviduct. The total number of eggs which a female bird may lay during her entire lifetime is already present, in the form of microscopically small cells, prior to reaching sexual maturity and before the first mating period. At mating time the appropriate number of egg cells for that season increases in size and forms a cluster which is, however, still very small. Only at this stage can they be fertilized. It is for this reason that fertilization does not occur when ducks, for example, play at mating in the autumn.

While the eggs of reptiles develop to depositable size simultaneously—thus producing a considerable weight increase in the mother—bird eggs develop only one at a time, albeit at amazingly short intervals. During its passage from the ovary through the oviduct to the cloaca, which normally requires a period of twenty-four hours, the weight and volume of the egg increases by several times. In the process, the yellow sphere of the yolk grows and is surrounded, in turn, by clear layers of albumen, by two thin elastic membranes,

and, finally, by a thin layer of calciferous paste which assumes the familiar smooth, ovoid shape while hardening.

The basic coloration of the egg is already present in a solution within the calciferous paste. This is a bilious dye and provides the egg with its basic green, blue, yellow, brown, or reddish pastel hues. However, the marking patterns which can be seen on most shells are not present until shortly before the egg is laid, and occasionally they can even be wiped off the freshly deposited egg. The spots, speckles, and splashes, the stains and delicate hairlines, which usually vary from dark reddish brown to almost black, are caused by blood clots in the oviduct. Consequently, one egg never bears exactly the same markings as another, although the type of pattern is often so characteristic for a given species that an expert can identify it without having seen either the bird or the nest.

The size of the egg in relation to the bird varies considerably. In general, birds which leave their nests early require relatively large shells, because they do not hatch until they have attained an advanced stage of development. But there are a great many exceptions.

The hummingbird lays the smallest egg: it weighs only approximately 0.25 gr and is about the size of a pea. The egg of the ostrich averages $5^1/_2$ inches in length, $3^1/_4$ lbs in weight, and its volume corresponds to that of about 25 chicken eggs. The egg of the gigantic aepyornis, the largest of all birds, which became extinct only very recently, was more than a foot long and weighed about $17^1/_2$ lbs, approximately as much as 180 chicken eggs or 32,000 hummingbird eggs. Shell fragments of these eggs are said to have been used as vessels by the natives of Madagascar as recently as hundred years ago.

Not every bird egg has the typical ovoid shape. Some owl eggs are almost spherical. Cormorant eggs are so elliptical that it is difficult to make out the pointed end. The eggs of nightjars are elongated and nearly cylindrical.

Prior to the breeding season, the effects of certain hormones on the thyroid gland produce the magnificent ornamental plumage of the great white egret; those long, delicately frayed decorative feathers on his back, which are raised high at courting time, to look like an airy white cloud.

Birds depend on sunlight. The light regulates the rhythm of their lives. Light stimuli received by the eyes are passed on to the brain, where they help to produce—in the pituitary body or hypophysis—the hormones: those miraculous messenger substances which arouse the sexual impulses, cause the ornamental plumage to sprout and schedule the mating and breeding moods.

Depending on the influence of light and on the position of the sun according to the season, the hypophysis will release larger or smaller quantities of hormones into the bloodstream, which then distributes them throughout the body, where they regulate and control the metabolism of the cells and organs, and particularly the hormone production of other glands which to a great extent affect the physical and sexual development of the bird. Thus, the sex hormones, which are produced in the hypophysis and whose concentration is increased by the lengthening days in spring, will stimulate the activity of the gonads. Other hormones affect the thyroid gland, which in its turn produces the hormones that affect moulting and migration.

THE DEVELOPMENT OF NEW LIFE

Weight economy is the top requirement in the conquest of the air, and an expectant bird mother cannot be too heavy to fly. For this reason, her brood must be produced by stages, one egg at a time, usually in a twenty-four-hour rhythm. Some larger birds ovulate every second day, a small minority at even greater intervals. But in every case, there is only one egg in the oviduct at a time.

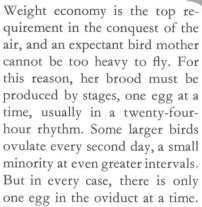

First day:
The yolk sphere floats in the albumen. The germinal spot, recognizable as a light, spot always floats to the top, as close to the heat as possible.

Third day:
The tiny red spot, the heart, is already beating before the embryo bears even the slightest resemblance to a vertebrate.

Sixth day:
The extremities are discernible. The black spot on the huge head can be made out quite clearly: the developing eye.

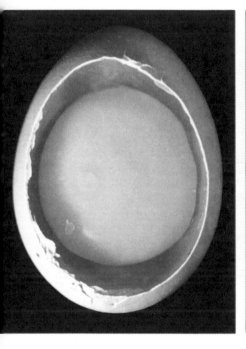
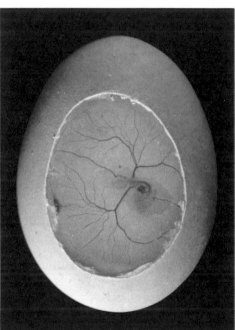
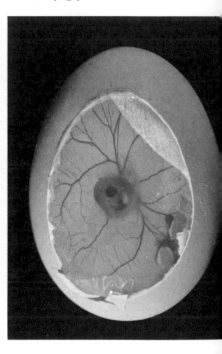

Sixteenth day:
The down covering has been formed in sticky strains. Each individual feather is covered with a thin protective film.

The development of new life has already begun before the egg is laid. The initial germ cell divisions, producing several hundred cells, occur inside the mother's body. But then, development is arrested until incubation begins. Heated to 104° F by contact with the skin, cell division sets in again, and now the young bird will develop in an unbelievably short time. The large spherical yolk always remains suspended in a central position, being held in place along the longitudinal axis by two firm albuminous bands called chalazae. The germinal spot rests on top of the yolk. It is lighter than the nutritive tissues and will rapidly float to the top again when the egg is turned, thus always remaining closest to the source of body heat. The embryo, a tiny, wormlike, transparent thing—hardly identifiable as a living creature if it were not for the clearly visible and constantly and vigorously pulsating heart—grows rapidly. Soon the huge head can be recognized, and the extremities can be made out a few days later, the down covering is formed on the fifteenth day, and on the twentieth day, the chick peeps inside the egg.

Zero hour: Follicular ejection. The egg separates from the egg cell cluster in the ovary.

After 6 hours: The yolk sphere is surrounded by albumen and by two thin membranes.

After 12 hours: The walls of the oviduct secrete the calciferous substance which forms the protective shell.

After 18 hours: In some species, camouflage markings on the shell are produced by blood clots in the oviduct.

After 24 hours: Ovulation. The second egg separates from the ovary at approximately the same time.

Tenth day:
The nutritive substances pass through the delicate vascular network of the yolk sac into the rapidly growing body of the bird.

Eleventh day:
The tiny bird already exhibits amazing vitality: from time to time it moves with vehement convulsions.

Thirteenth day:
The yolk sac is clearly vanishing. The eye is still dominant, but many details, such as toes, are well developed.

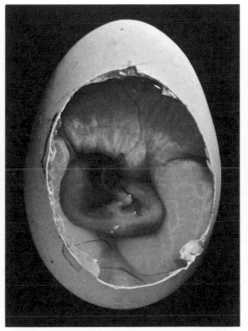

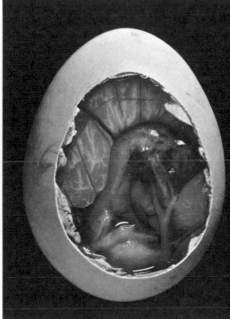

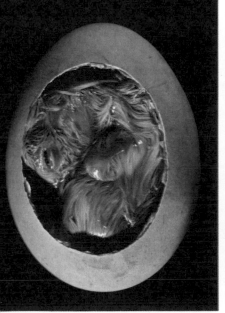

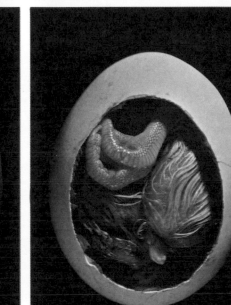

Twentieth day:
The chick is fully developed. And now a most important extremity of an early fledgling is dominant: the foot.

Twenty-one days are required for the development of the domestic chicken embryo illustrated here. According to the species, the incubation period may last from ten and a half to eighty days. The two following pages illustrate how the chick emerges from the egg.

1   6 p.m. The first tiny cracks become visible in the shell of one of the coot eggs.

2   7 a.m. By the following morning, the coot chick has pecked a hole through the shell with its sharp egg tooth.

3   11 a.m. The chick has weakened the shell with several small holes and has made a larger opening in the vicinity of its head.

4   2:30 p.m. By now it has used its head to pry off the blunt end of the eggshell like a lid.

5   2:50 p.m. With utmost exertion, repeatedly interrupted by long recuperative pauses, the chick struggles to extricate itself from the shell.

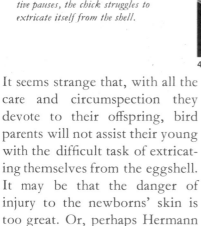

It seems strange that, with all the care and circumspection they devote to their offspring, bird parents will not assist their young with the difficult task of extricating themselves from the eggshell. It may be that the danger of injury to the newborns' skin is too great. Or, perhaps Hermann Hesse's dictum might apply: "He who would be born must destroy a world."

On about the twenty-first day of incubation the coot chick will open the air chamber, enclosed by the inner lining of the shell and located near the blunt end of the egg, with its bill. From then on the creature can breath and thus, for the first time, it can make its presence known to the outside world: it begins to make peeping noises from within its "prison." Among the larger birds these sounds can be heard quite distinctly from a distance of up to about six and a half feet. The parents of some species will reply to these calls, thus establishing the first parent-child contact. Approximately two days later the chick breaks the first tiny hole through the shell with its sharp egg tooth. After a long pause a second hole becomes visible close to the first one. Gradually, the chick will punch several small openings into the shell, often puncturing the whole circumference of the egg. Whenever it pushes vigorously against the shell, the chick's body twists slightly in the opposite direction, frequently enlarging the hole surrounding the bill. Through the various openings the chick can soon be observed making determined intermittent attempts to stretch itself, simultaneously pushing vehemently with its bent neck against the blunt end of the egg until it snaps off like a lid. Again the chick must recuperate for a long time before it struggles out of the egg by stages, a laborious process which may take anywhere up to an hour. During this period it also begins to straighten itself out from its curved embryonic position.

The newly hatched chick is not really as wet as it appears. Each individual down feather is enclosed by a delicate protective covering—a clear albumen, which makes them resemble fine, wet, wispy hairs. But the hair-shaped down feathers themselves are kept completely dry in their protective sheaths. After a short

The bird
fights to emerge
from the egg.
It is his world.
He who would be born
must destroy
a world.

HERMANN HESSE

6 3 p.m. Almost completely emerged from the shell, the chick stretches its neck for short moments, but it still retains the embryonic position.

7 3:20 p.m. After it is finally released from the shell, the curved embryonic position is gradually abandoned.

6

time the albumen dries off, and as the chick begins to move about the nest, brushing against the edge of the nest, the eggshells, its brothers and sisters, and its parents, the delicate covering crumbles, the downs unfold and the ugly little bundle is transformed into a charming, fluffy chick.

The entire hatching process, beginning with the appearance of the first hole, requires about twenty-four hours for bird species approximating chickens in size. Smaller birds manage to hatch in a few hours, while very large ones may require up to three days.

7

On the following pages: 4 p.m. One of the parents has removed the telltale eggshell from the nest, and the newly hatched chick is almost dry.

The cuckoo has long been a symbol of good fortune and long life on the one hand, and of indolence, parasitism, and infanticide on the other. To the European the familiar bisyllabic call of the cuckoo is a welcome harbinger of spring. The bird's behavior, however, is less agree-

able; the cuckoo smuggles its eggs into alien nests and leaves the arduous task of raising its young to others. We find it difficult to accept the idea that the habits of the cuckoo are beyond the concepts of good or evil; that its behavior is largely instinctive and in accord with the characteristics of its species.

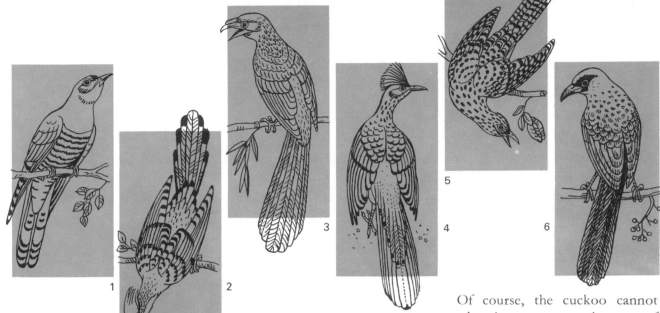

1 *The cuckoo practices parasitic breeding in Europe, Africa, and Asia.*

2 *The great spotted cuckoo of Southwest and Southeast Europe and Africa lays its eggs in crows' nests.*

3 *Several greater ani females usually lay their eggs in a single nest and all join in hatching them. This species lives in South and Central America.*

4 *The road runner hatches its own eggs. It is one of the long-legged cuckoos that live in South and Central America and in the Southwest of the United States.*

5 *As early as the year 375 B.C., those koels of Asia and Australia were known to lay their eggs in strange nests.*

6 *Coucals, which live in Africa, South Asia, and North Australia, do not practice parasitic breeding.*

The unsuspecting reed warbler feeds the young cuckoo. The red, gaping bill of the "changeling" seems to hold an irresistible fascination for the foster parents. They will feed it just as devotedly as they would look after their own offspring.

The cuckoo family is made up of about 140 species. Almost 65 percent of these look after their own offspring. The remainder are partly or exclusively parasitic breeders, but nearly all restrict their activities to a few—often only one—carefully selected host species.

The European cuckoo is the most notorious of all parasitic breeders, and for good reason, for it has achieved unparalleled mastery of this dubious talent. Its solution to the propagation problem renders matrimonial ties superfluous. The males do have their own breeding areas, which they defend and retain, but the females roam about freely, permitting themselves to be courted and seduced indiscriminately.

Most birds lay their eggs early in the morning. To avoid clashes with the intended hosts, however, the female cuckoo does not lay her egg until afternoon, and will, wherever possible, select a nest which does not yet contain its full complement of eggs. She will loiter clandestinely in the vicinity of the nest and will approach it as soon as it is left unguarded for a moment. Within a matter of seconds she will deposit her egg in the nest and then pick up one of the host's eggs in her bill and disappear with it. The size, shape, coloration, and markings of the cuckoo's egg must conform rather precisely to the eggs of the host.

Of course, the cuckoo cannot adapt its egg to any given set of eggs it may happen to come across (as has been claimed on occasion). The female cuckoo, like all female birds, produces but one type of egg throughout her lifetime—the same that was produced by her forebears and will be produced by her offspring. This egg matches that of only one, or in some cases a very few, host species.

More than twenty different bird species serve as foster parents for the various types of cuckoos. If the host bird does not notice the swindle, the young cuckoo will hatch after the extremely short incubation period of ten and a half days; i.e., before his stepbrothers and stepsisters, who normally require from twelve to fifteen days to hatch. Naked, blind, and ugly, the young cuckoo conveys the same impression of helplessness as any other newly hatched nestling, but the impression is deceptive. Only about ten hours after it has hatched it begins to slide nervously about in the nest until it comes in contact with a solid object—that is, one of its unhatched fellow occupants. It will wedge itself underneath the egg, rear end

The examples on these pages show how accurately the various cuckoos' eggs can match the eggs of various hosts. In each case the cuckoo's egg is shown on the right.

Jungle babbler —Common hawk cuckoo

Robin—Cuckoo

Red-backed shrike—Cuckoo

Whitethroat—Cuckoo

Brown hill warbler —Cuckoo

first, grasping it firmly with its little wing stubs and then, summoning all its strength, it will push the egg backward over the edge of the nest. The baby cuckoo will continue in this manner until the nest is empty—

Garden warbler—Cuckoo

its "expulsion instinct" remains active for several days.

To sustain its enormous rate of growth, the young cuckoo needs all the food which would have sufficed for an entire brood.

Magpie —Great spotted cuckoo

During the very first days of her life the female cuckoo receives lasting impressions of her foster parents; impressions that later will help her to identify the nests in which she must lay her eggs as an adult, and to know which eggs will match her own. This bond with her foster parents, however, concerns only their nest. Hence,

to deposit her eggs, she will seek out nests of that same or of similar species whenever possible.
In this way, numerous so-called "biological cuckoo races" have come into existence in the course of time: reed warbler cuckoos, white wagtail cuckoos, robin cuckoos, etc. They all look alike, of course, but each "race" has its

standard hosts as well as several alternate hosts. A female cuckoo rarely will visit an "emergency host," and if she does, her visit usually ends in disaster when the fraud is discovered and the alien egg is rejected.
While the cuckoo is the undisputed king of parasitic breeders and has developed unequaled

Chinese thrush—Red-winged Indian cuckoo

Blackcap—Cuckoo

Reed warbler—Cuckoo

Large spiderhunter
—Large hawk cuckoo

Marsh warbler
—Cuckoo

Yellow-bellied prinia
—Little cuckoo

the cowbird among American blackbirds, and, in particular, the black-headed duck, which lays its eggs exclusively in alien nests. In the latter case, the nests usually belong to other ducks, but occasionally the parasite will seek out even the aeries of birds of prey. Almost all the aforementioned birds will normally entrust only close relatives with their eggs; something the cuckoo never does. And most limit themselves to standard hosts of a single species and produce only a single type of egg which happens to match the eggs of the host.

The "expulsion instinct" of the young is also known exclusively among cuckoos. The African honey guides, however, have developed a most interesting parallel technique. As is the case with the cuckoos, the chicks of these woodpeckers invariably are encountered as singletons in the nests of their hosts, and while still naked and blind, they already have become killers. At the tips of their beaks they have two pointed hooks with which they literally mangle their stepbrothers and stepsisters. About ten days later, when these instruments have served their murderous purpose, the two hooks drop off.

mastery of its technique, it is by no means the only bird which leaves the care of its offspring to others. No fewer than 82 species, belonging to the most diverse families, are known to multiply exclusively by means of such practices. Furthermore there are many additional species that occasionally practice parasitic breed-

ing. Parasitic breeders, incidentally, also occur among insects; among a number of bee species, for example, characteristically known as "cuckoo bees."

Instances of "unscrupulous" parasitic breeders are the honey guides among the woodpeckers, the cuckoo weaver and all whidah species among the weaverbirds,

WIVES SWEETHEARTS CONCUBINES

In the course of their development, birds have entered into and experimented with countless forms of connubial life—forms which can be found in one or another species to this day. To be sure, they have refined the bonds of matrimony step by step, but, as a result of their specialization, the "perfect marriage," the permanent commitment of a couple for life or at least for one breeding season, has not proved to be the only acceptable mating pattern. Thus we find, even today, among the multifarious forms of bird marriages, inveterate monogamists as well as ardent polygamists.

The development of the marriage commitment is probably closely allied to the nesting instinct, and serves the same purpose, namely, care for the young and protection against a hostile environment.
The females of ground breeders, whose young leave the nest early, can generally function without the presence of their husbands. They do not have the laborious task of feeding each individual young; instead, they lead their brood, which is quite independent even at birth, to the food. Thus most female gallinaceous

Their profound dislike
of
monotony,
their unbridled delight
in variation
does not
stop short
at the state
of matrimony.

birds and ducks manage to raise their offspring on their own. Even among these species, however, there are numerous exceptions. For example, male rails and lapwings are solicitous fathers who help their partners in leading the chicks when they leave the nest. Moreover, cranes and geese, who also breed on the ground, are not satisfied to marry for only a season, but enter into wedlock for life. Still, there are numerous ground breeders who demon-

protective treetops can normally remain confidently in their nests until their wings are able to support them to some extent. Their cradle is usually so well concealed that they are in no great danger of nest robbers. This is also true of troglodytic breeders, woodpeckers, titmice, and of all the other numerous species whose nurseries are hidden in hollow trees and in rock or earth caves. But the increased security of these sheltered nests places other

Here too, there are a number of exceptions. The female birds of paradise, for instance, take care of brooding and of raising the young on their own, although their offspring stay in the treetop nest until they are fledged. There simply are no hard and fast rules. But in general, the bird species whose young remain in the nest contract longer lasting marriages than those whose young leave the nest early.

strate that life can be quite pleasant without a long-term marriage.

The majority of birds prefer the arboreal life because it offers greater security. Young birds which are hatched among the

demands on the parents: the food must be procured and brought to the children, and in rather considerable quantities. Thus it is of decided advantage to the maintenance of the family that the father should remain to lend his assistance.

A number of birds favor what we humans consider to be the "perfect marriage:" the commitment which lasts until the death of one of the partners.

The young of most bird species lead a restless life as members of flocks, frequently undertaking migrations that lead them far beyond their usual habitats. But when they reach sexual maturity, they look for a suitable domicile in an area as nearby as possible like the one in which they were born. Now, the mating games of their youth, which hitherto had been carefree (and also clumsy), take on a more urgent and precise character. Couples are formed one after another. With the start of nest construction they will have joined in a bond for life. They quickly become so devoted, so accustomed to each other, that they will stay together even after the first brood has matured and departed. Frequently, such a couple will remain alone in the nest area until the next breeding season sets in, and will rout every intruder of their own species. Given such tendencies, the startling fact is that the majority of such monogamists, like the polygamous species, will assemble in the autumn to mi-grate in great flights to their winter quarters in the south, still retaining their identities as couples. As we have seen, some birds of prey—particularly the larger species—remain together as couples in the same area throughout the year. Even the transients do not necessarily assemble in flights, but often migrate as couples. Barn owl couples will remain in one area for years, as do the ravens. Geese, storks and cranes, on the other hand, often migrate in large communities to their winter quarters, where they

continue to lead a gregarious life. Among some species, the couples even separate for a certain time, and in these cases it is probably loyalty to a locality, not the partner, that eventually reunites the couple at their old accustomed breeding ground. Such is the case among white storks. They return from their winter quarters as individuals and rendezvous at the nesting site. There, if a younger female attempts to enter an aerie accupied by an older couple, the male will calmly watch his mate try to rout the intruder. Whichever of the contending females proves herself the stronger, is the one with whom he will celebrate his nuptials, and it appears to matter little to him which one it will be.

Left: African sea eagles make excellent marriage partners and remain true to each other until one dies. The couple will always remain in the same area and keep in constant touch by means of the shrill, savage, wonderful calls that have earned these birds the epithet, "the voice of Africa."

Right: Scarlet macaws and military macaws at an earth slide on the bank of the Rio Manu in Southeastern Peru, where the birds are eating the loamy, salty earth. This photograph is probably the first ever made of macaws on the open range.
Parrots generally are very gregarious birds. They often roam together in great flights in search of feeding grounds. Yet, close observation will show that the individual couples, especially among the larger species, do not stray far apart during the flights. When a couple is ready to mate, it will usually leave the flight and look for a suitable hollow in a tree. Some species breed in more or less loosely associated colonies, and the green parakeet even breeds in the close company of an apartment building made of brush.

Long Engagement – Brief Marriage

A strange type of marriage seems to be preferred by most ducks. The drakes, as the male ducks are called, are particularly fine, extremely solicitous marriage partners. They do not wait until breeding time to sue for the favors of the females, as most birds generally do. Instead, they court them in autumn, and when they are accepted they remain faithfully beside their plain, unadorned fiancées throughout the winter. But in spring, when the marriage has been consummated at last at the end of this long "engagement," and the duck has bedded her numerous eggs in the downy lining of the nest and faces the imminent task of hatching and raising the offspring, the seductive drake reveals himself for the windbag he really is. The duties of a father are not to his liking. He dissolves the relationship in short order and joins the carefree company of other drakes. While the perfidious males make a merry summer of it, the abandoned females patiently hatch their eggs and lead their precocious chicks to favorable, sheltered feeding grounds.

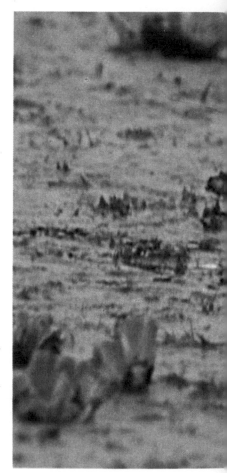

To be sure, most drakes' magnificently colored plumage would ill suit them for brooding and leading the young. They would be noticed far too readily by egg robbers and chick murderers, while the plain plumage of the female offers excellent camouflage. The sitting, immobile duck blends marvelously with her environment, and she seems to be fully aware of this fact. Even at extremely short range she will usually escape detection, and thus she will remain on her eggs until one is almost close enough to touch her.

That ducks are rather critical in their choice of partners and will not accept the first one to come along may account for their long "engagement" period. In autumn, they assemble in large flocks to seek their fortunes, and soon begin a frolicsome and even licentious community courtship. One female, for instance, head lowered and nodding vigorously, may suddenly swim about in curves and loops among numerous males. Or the males may start

period of "normal" life and rest.

This autumnal community game is apparently of great importance for the brief marriage of ducks. The conspicuous behavior continuously attracts additional passing ducks. The larger the flock, the better the chance that even the most exacting individuals in it will find a suitable partner. And since the game does take place in autumn, there is no reason for hasty action. There is ample

conclusion that the males are after a still-single duck, and that each of them is trying with might and main to win her for himself.

At this time, however, the marriages long have been contracted, and the drakes merely are having an inconclusive flirtation while their wives are looking after the home. Quite often a drake will plan an escapade and unexpectedly give someone else's wife a rush. Lady ducks being modest,

shaking and twisting as a group, raising their tails and bending their necks backward. The whole company gradually will be caught up in an ever-increasing pitch of excitement, almost to the point of ecstasy, which will break off abruptly, to be followed by a

leisure to shop around before making a commitment. In the spring, two drakes can be seen in flight now and then, as they pursue a female duck which desperately seeks to evade their importunities. Such a sight might lead an observer to the natural

she will immediately take flight with the "rake" in pursuit—and with the rightful husband at *his* tail. After a brief, wild chase the would-be seducer will abandon his risky, hopeless enterprise and make a bee-line back to his patient spouse.

Seasonal Partnerships

2   1   3

4   5   6

Marriage lasting for a single breeding season seems best suited to the life of most birds and is the most frequently praticed form of bird matrimony. To be sure, there are many other kinds of partnerships. Many more birds than is commonly believed, for example, remain with the same partner for years. But who can claim intimate knowledge of the "love life" of all the vivacious, feathered species? Be that as it may, among most species the marriage partners share a very intimate relationship. Observing a pair of birds greeting each other exuberantly as one relieves the other of the brooding task, caressing each other, or nestling against one another while enjoying a rest, it is difficult to avoid the impression that the couple is deeply in love—in the human sense of this expression—and that the relationship is far more cordial than would be required by the mere care for the offspring. Even bird couples which maintain less friendly matrimonial relationships than are customary among most feathered species share in brooding and raising the young. The woodpeckers are a case in point. With very few exceptions, these "choppers" are confirmed loners.

7   A common tern brings food for the little ones. When terns are courting, they often will demonstrate their affection with presents of small fish.

8   Marital relations between woodpeckers are tense. This red-bellied woodpecker also has a tale to tell.

9   A pair of rare birds: roseate terns.

10   This pair of Galapagos penguins is rarer still: there are only about two hundred of these birds alive today.

11   Once the frigate birds have established their partnership, the red balloon lure of the male is deflated.

On the following pages:
A couple of swallow-tailed gulls greet each other with the raucous noises customary among gulls.

7

8

9

Outside the breeding season, when most birds become especially gregarious and assemble in flocks, each woodpecker will continue to reside in its own area. When two of these hermits meet, they threaten each other with fearsome calls, peculiar thrusting feints with their bills, and aggressive flying maneuvers. It may eventually result in a bloody fight, and, since the beak of a woodpecker is a dangerous weapon, such a quarrel may end with the dead of one of the birds. Even the mating games of woodpeckers are far from being tender declarations of love, and would be more aptly described as "menacing courtships." The future marriage partners face each other in threatening attitudes. Neither will let the other approach too

10

11

closely, each establishing the appropriate distance with repeated feints of the beak. Both seem to be extremely hesitant about entering into a relationship, even for procreation and the establishment of a family. Once the eggs are laid within the chip-lined nesting cavity, however, the parents will brood alternately, the male sitting at night, as a rule.

When the partner approaches the tree for the brooding relief, there is no exchange of the tender greetings customary among other birds; the partner who is relieved of duty departs from the nest almost as if in rout. This rather hostile relationship even includes repeated biting bouts, and it is dissolved without delay as soon as the young are fledged.

125

The male ostrich conquers
his harem
with an ecstatic dance.
But soon he proves
that he is more than an ardent beau:

The African ostrich practices what is probably the most peculiar form of matrimony in the entire bird kingdom. With low rumbling calls and odd contortions of his neck, trembling all over and wildly flapping his wings, the gigantic male ostrich will pursue first one hen, then another, and yet another (his harem usually consists of three wives). Then he will settle in a pit which he has dug in the sandy ground during the courtship. Soon, each of his hens will deposit four to ten eggs at two-day intervals in front of his breast and he will carefully push the eggs under his belly with his bill, himself assuming the major responsibility for the brooding. He is relieved in this duty temporarily, mainly during the daytime, but only one of the hens, his favorite, does the relieving, for his is a very formalized type of polygamy. The other wives are permitted to deposit their eggs in the nest, after which they are driven away by the favorite. The size of his harem may vary, but the wise ostrich will limit the number of his wives. If the harem were too big, there might be too many eggs for incubation, with the result that no chicks would hatch at all.

Left: The peculiar form of matrimony practiced by the ostrich is shown schematically in this illustration; the cock usually has two concubines in addition to his favorite female.

dutifully he settles on

the heap

of giant ostrich eggs.

Only his favorite

may participate in the parental care.

Below: The cock permits only his favorite to assist him in brooding the eggs of his harem and in guiding the chicks. Here she searches for food along with several adolescent youngsters, while the cock remains in the vicinity, looking after the remaining little ones.

The Married Life of Fowl

Fowl-like birds are—for the most part—inveterate polygamists. Consequently, the mating habits of that rare bird, the monogamous fowl, are particularly intriguing. The marriage of the rock partridge is an example of this rarity. After mating, according to the ornithologist Heinz-Sigurd Raethel, the female of these one-season monogamists will scratch a trough into the ground and fill it with about eight to ten eggs. Then she will dig a second trough—usually about one hun-

duration of one breeding period. To be sure, they do not share the parental duties as equally as the

Glimmering
under glassy skies
lies the waste
of the semi-desert,
hostile to life.
Game tormented
by the scorching heat
stands in the
filigreed shade
of gray bushes,
brownish yellow
like the dust
and the
burned-out vegetation.
Then
like a steel-blue gleam
amidst
the dry-brown grass
a flock of guinea fowl
relieves
the starkness
of the barren plain.

Above: The rock partridge couple hatches two sets of eggs simultaneously: the hen cares for one nest, the cock for the other.

Right: In the northern part of East Africa, one can occasionally find vulturine guinea fowl living in the semi-deserts in huge flocks. Throughout the breeding season they stick together in pairs.

dred yards away—and deposit in it approximately the same number of eggs as in the first. Then something quite unusual takes place: the hen will sit on one of the two nests, while the cock will look after and brood on the other. As soon as the young are hatched, the couple reunites and together they care for their large flock of chicks. Peacocks, on the contrary, are Don Juans par excellence—they consider their parental duties to consist of copulation with as many hens as possible. But their close relatives, the guinea fowl, are monogamous, at least for the

rock partridge. The males do not concern themselves with the eggs; still, they do remain in the vicin- ity of their brooding partners, and when the young hatch, they assist in raising them.

The rooster has come to be the symbol of fertility and of untempered passion—the Don Juan of the feathered kingdom. Most of his relatives are polygamists whose parental duties are done when they have fertilized as many hens as possible.

Guinea fowl live as monogamous couples within their flock, as shown in the schematic representa- tion here. The cock always remains near the sit- ting hen and assists her later in raising the chicks.

With few exceptions, grouse cocks court their hens in groups.

Curassows are monogamous; the larger species probably live in this manner for many years.

Hoatzins are monogamous, but they nest in communities in bushes and low trees.

Also within the grouse family, which includes such passionate polygamists as the black grouse, the capercaillie, and the heath hen, there are species—namely the hazel hen and the ptarmigan—who live in exemplary monogamy. The male of these species also remains near the brooding hen as a rule, and later helps her in guiding and warming the chicks.

The oldest known fowl skeletons were found in geological formations approximately fifty million years old. But even without such fossil evidence, the gallinaceous order can be recognized as an ancient category of birds. They have remained, for the most part, ground dwellers; they stand on sturdy feet, and in the face of danger many of them seek concealment among low vegetation or escape on foot rather than taking wing. They also have retained a more or less homogenous external appearance, despite considerable variations in body length of various species. And lastly, most of the 263 known species of fowl persist in an ancient form of bird mating—the practice of promiscuous polyg-

amy. The fact that among these families there are several exceptions to the rule—such as the guinea fowl, the rock partridge and the ptarmigan mentioned above, as well as the curassow and the hoatzin, all of which practice monogamy to some extent—evidently has failed to impress the bulk of their relatives.

In the mating season the cocks almost invariably lure the hens with loud, ostentatious calls, while they stand, each in his own territory, defending their ground against possible intruders. Among some species, several cocks may assemble for joint courting, staging impressive mock duels, as if to demonstrate their potency to the hens. Then these magnificent fellows will proudly display their iridescent plumage, putting on airs and graces that give the impression of extreme vanity. But the mating itself usually lasts only a few minutes: when the enticed hen has been fertilized, the cock no longer bothers with her. He returns to his social life—sometimes among huge flocks.

In most cases the hen wears a plain camouflage plumage—as do ducks and many other ground breeders—which affords her ex-

cellent protection as she sits quietly on her eggs. On the other hand, the magnificent dress worn by some cocks all year long ranks among the most beautiful of all bird plumages. Especially during courtship, when they parade and display themselves with particular skill, the finery of some species surpasses even that of the birds of paradise. The colorful splendor and picturesque forms of court-

When there are not enough wild turkey cocks to congregate in the usual male flocks, as a result, perhaps, of decimation, they will join groups of

Cock pheasants lure as many hens as possible, but, after mating, the hens must tend their brood on their own. Some species, however, are monogamous.

Mound fowl are also polygamous. The cocks of most species, however, take care of the eggs, which are hatched in nests like incubating ovens.

Turkeys, like most pheasants, are polygamous, but several hens will frequently brood in close proximity to each other; occasionally two or three will share a nest and pool their eggs during incubation.

ing peacocks, golden pheasants, and tragopans can hardly be outdone!

Wild turkeys are rather typical fowl in their mating habits. Their home is in the sparsely wooded areas of the southern United States, but they have been decimated in large sections of their original habitat.

Wild cocks—in contrast to the

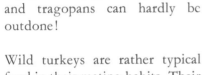

hens and young birds. The illustration below shows three wild turkey hens and two young birds, led by a cock.

domestic turkey cocks which can produce offspring after one year —normally do not reach sexual maturity until their second year. Then, when they are ready for mating in the spring, they occupy their individual courting areas and begin to "gobble," that is, to voice their mating call, which sounds like "gobble-gobble-gobble" (and which has earned the male turkey the name "gobbler" in America). At the same time they will spread their contour feathers, shake their drooping wings and fan their tail feathers, thus giving themselves a larger and more impressive appearance. The bare, warty skin areas about their heads and necks swell and take on a bright red and blue coloration. When a hen approaches, lured by his calls, the cock appears to pay no attention to her whatsoever. He will affect utter disdain, and the hen must literally prostrate herself in front of him before he will condescend to engage in copulation. This blasé attitude on the part of the male is, however, deceiving. In reality, the hen holds his complete attention—even when he repeatedly turns away from her in the course of the courting ritual. For when a rival attempts to contend

for his area, the turkey cock shows his true desire for his prospective mate. He then becomes extremely excited and will try to make himself appear as awe-inspiring as possible. At times the intruder will nevertheless risk a fight, a serious damaging duel of the kind that seldom occurs among animals of the same species. In such a fight one of the cocks may even be killed by the other. But as a rule the intruder will quickly acknowledge defeat by throwing himself flat on the ground in an attitude of humility. Thus he is spared from further attacks.

After mating, the hen leaves her partner, scratches a hollow into the ground among the underbrush, lines it with plant debris, and deposits from eight to twenty eggs. When she leaves the nest in search of food, she covers the eggs with foliage. In fourteen days after hatching, the chicks can already fly, and will follow their mother into the treetops to spend the night there. Only during the actual incubation periods will turkeys sleep on the ground; at other times they prefer high roosting sites, like most other woodland fowl.

FAITHFUL BEYOND DEATH

The married life of geese is one of the most "exemplary" forms of monogamy. Whether they brood in crowded communities —like the Canada geese shown here—or whether the couples isolate themselves at mating time, they usually remain true to each other for life—and even after one partner dies the other rarely takes a new mate.

Geese remain faithful to each other for many years, usually until one dies (and even then the surviving partner usually does not mate again), though many species brood gregariously in loose colonies and often assemble in huge flocks outside of the breeding season.

The mating of geese, like that of ducks, their close relatives, is a long process involving complicated rituals. But individual couples will simplify the introductory courting gestures somewhat from year to year—and an older couple which has become attuned to each other over the course of many years may give them up almost completely, as though they had forgotten them. When an elderly goose dies, the widowed partner is no longer capable of performing the elaborate courting gestures of its youth and hence cannot attract another, younger partner. Thus "faithfulness beyond death" is actually a consequence of "forgetfulness" and should not be ascribed to mourning over a painful loss.

While male ducks, or drakes, often leave their mates "in the lurch" when it comes to hatching and raising, the young male geese, or ganders, will keep watch over their brooding partners and then share in herding and warming the young. Male black swans which also are relatives of the geese—and the males of various tree duck species—even help with the incubation process.

Canada geese are a particularly sociable species. They like to live in communities which are as large as possible, and when their goslings are a few weeks old, they herd them collectively. But even in great flocks, the individual couples will be faithful to each other year after year.

Male birds whose resplendent colors too flamboyantly upstage the inconspicuously plumaged females of their species are definitely suspect of polygamy. Few of these decorative males will continue even to pay attention to their partners once the impetuous mating is over, let alone burden themselves with parental duties. Almost invariably they are insatiable Don Juans who rush from one partner to the next. Male ducks behave far more "decently" by comparison: they will leave only one female in the lurch after mating.

It is odd not only that these "lady-killers" exist in several orders of birds, but that faithful males may be found alongside ardent polygamists even among closely related species. It is true that polygamy is practiced more frequently among birds which are older in terms of their evolutionary history, but there are also some "unscrupulous" seducers among the younger evolutionary orders. For example, male whidahs—members of the finch family—desert the females after mating. The females, however, can fend for themselves. They have no taste for building a nest, brooding, and raising the young on their own, and so they foist off their eggs on other birds, specifically on certain waxbill species—one might, indeed, be justified in calling them "Merry Whidahs."

Don Juans exist among crows as well, although the great majority of male crows make very good fathers indeed and will live with a single partner for at least one summer and sometimes for life. They will even help the female to brood and are indefatigable in gathering food for their young. But their relatives, the birds of paradise, have absolutely no interest in monogamy and the care of the brood.

Wren

Even those conservative male birds, whose plumages are almost indistinguishable from the females, however, do not necessarily always make the best heads of families. The tiny wren, for example, generally builds a whole series of mock nests. At breeding time, he will use one of them to seduce the first female that heeds his song, which he warbles with amazing volume. By the time she has padded the nest for her purposes, moved into it, and begun to brood, her lord and master is already off looking for another playmate who will exchange her love for one of his pretty little spherical houses.

The ruff is a most amazing bird in several respects. He is one of the sandpipers, a group of waders

Bittern

Paradise widowbird

Ruff

Golden pheasant

Superb lyre-bird

Lesser bird of paradise

Golden-crested bowerbird

which range in size between a sparrow and a pigeon, whose sexes are identical in coloration and which usually practice monogamy (the male of the Temminck's stint will even hatch one of the two broods produced in rapid succession by his mate). But the ruff is not much of a family man. Like many gallinaceous birds, he will meet in the spring with a number of other males in swamp meadows at tournament sites which have been used for display purposes repeatedly over a number of years. There he will display his plumage with its great erectile ruff which is unique among shore birds: it contains a multitude of hues ranging from rusty red to black and white, and the color combinations vary considerably among individual birds. Arrayed in this manner, these splendid fellows display themselves in ferocious-looking mock battles. In the pauses between these turbulent bouts, the females, or reeves will walk up to the crouching males and invite them to mate by pecking at the ruff on their necks. This passive behavior on the part of the ruffs may be one of the reasons that the reeve, like the ruff, does not necessarily restrict herself to only one mate.

The Weaverbird: Ace Architect and Heartbreaker

Many male weaverbirds are Don Juans—but at least they put an excellent and nearly finished nest at the disposal of every female they seduce. Most species brood in colonies, which often consist of more than two hundred nests.

We have already come to know the wren as an avid amorist—as well as a rather hard worker, since he provides each of his conquests with a nest that is ready for occupancy. Well, many weaverbird species do almost the same thing.

Looking at the slovenly hay nests of house sparrows, it is difficult to imagine that they were built by close relatives of the finest architects of all birds—the weaverbirds. In other respects, of course, these relatives have much in common. Many weaverbirds, like sparrows, are proper "gadabouts"; they flit about the countryside in noisy, pillaging gangs and will settle unceremoniously right in the middle of villages. With the same mixture of prudence and impudence, they go about collecting garbage and pilfering unprotected food supplies.

Their gregariousness causes weaverbirds to crowd together even at mating time. Some establish their breeding colonies in the branches of trees or on the roofs or porches of houses, while others breed among reeds or papyrus. As many as two hundred or more nests may be built in close proximity to each other—a situation that naturally makes for some rather clamorous activity.

The construction of the ingenious weaverbird nest is begun by the male. First he will braid a perpendicular ring of grass blades and plant fibers; then he will weave a bag-shaped brooding chamber on one side, closing up the other side except for a round opening near the bottom, to which some species will attach a braided entrance tube of varying length. When the raw construction is finished, the architect will dangle, head down, from the nest whenever a female approaches, enticing her by chirping and beating his wings. If she is still unattached and available, she will inspect the nest, and usually quickly give her consent to the match. The soft interior finish she will have to provide for herself, however, for her short-term mate will not be bothered with further decoration—nor with brooding and feeding. He will use that time to build a new nest—to entice his next mate.

THE ORDERS OF BIRDS

This classification is based on Grzimek's Tierleben, vol. 7–9 (1969). A modification of Wetmore's classification, which, at present, is the one accepted by most ornithologists.

ORDERS	NUMBER OF SPECIES		ORDERS	NUMBER OF SPECIES
TINAMOUS	43		LARIDINE BIRDS	334
RATITES	10		PIGEONS	318
GREBES	9		PARROTS	326
LOONS	4		CUCULIFORM BIRDS	146
PENGUINS	18		OWLS	144
TUBENOSES	92		NIGHTJARS	96
PELICANS AND THEIR ALLIES	54		SWIFTS	77
WADERS	115		HUMMINGBIRDS	321
FLAMINGOS	5		MOUSE BIRDS	6
WATERFOWL AND SCREAMERS	151		TROGONS	34
DIURNAL BIRDS OF PREY	291		ROLLERS	190
GALLINACEOUS BIRDS	263		SCANSORIAL BIRDS	383
CRANES	199		PASSERINE BIRDS	5118

MONOGAMY—GREEN

The most general form of bird union is monogamy. Among all orders of birds there are at least a few species practicing monogamy, if only of short duration. The tinamous are perhaps the only exception—while there are a small number of monogamous male tinamous, they will find their mates among the polygamous females.

POLYGAMY—RED

The most common form of union among gallinaceous birds is polygamy, which is also practiced by approximately one third of the other orders. These partnerships are usually brief and are dissolved before ovulation. Polygamy is also practiced among most of the polyandrous species.

ABBREVIATED UNION—YELLOW

In many cases this can hardly be regarded as a partnership, since the male and female are together only during the courtship ritual and copulation and will separate immediately thereafter. While many ducks mate as early as autumn, copulation will not take place until the spring, and the couple will separate at the time of ovulation.

UNION FOR A SINGLE BREEDING SEASON—BLUE

The monogamous partnership that lasts for a single season is the most popular form of mating. Tinamous are probably the only order among which monogamy is not practiced at all, and among ratites it is practiced only by the ostrich in a peculiar variant of polygamy involving a single favorite wife.

UNION FOR LIFE OR FOR MANY YEARS—WHITE

It is very difficult to make an accurate determination of the species which mate for life, particularly among small birds, gregarious breeders, and the many species which form large flights out of the breeding season. Consequently, this form of union may be more widespread than is presently known.

PECULIAR BREEDING PRACTICES

POLYANDRY

Polygamy is, as we have seen, a rather widespread form of mating among birds. Polyandry is also practiced in a number of cases. But neither can properly be called a partnership, because the union generally lasts only for a very short time and, in most cases, is limited to the courting ritual and copulation.
Female promiscuity, however, is not practiced unilaterally, as far as we know. The males of the species are not satisfied with one mate either. Thus, a female cuckoo will permit herself to be seduced by several different males, and the male, for his part, will seduce every female he can entice into his area!

Polyandry is also practiced among tinamous, an odd, rather ancient order of birds which are related to the ratites, such as the ostrich and the American rhea. Because of their anatomical peculiarities they have been conceded a place of their own which ranks first in the evolutionary order of birds, even before the ratites, since they are considered to be one of the latter's ancestral groups. Tinamous are, incidentally, unlike their nearest descendants, capable of flight—though they are no experts at it. This is not particularly surprising; it has been known for some time that the flightless ratites are descended from birds which were able to fly, just as the ancestors of the penguins could fly.
The mating pattern of tinamous also is similar to the ratites, in that the brooding is also done by the males. To be sure, there are exceptions to this rule among ratites—the ostrich, for example, generally has brooding assistance from his favorite wife—but among emus, cassowaries, and American rheas the care of the eggs and the offspring is left exclusively to the males.
When a tinamou cock is in the mating mood, he will entice hens into his area with his calls (most tinamous have a soft, melodic call, often quite melancholic—not at all like the raucous mating calls we are accustomed to hear among gallinaceous birds). An occasional hen will approach and, with raised tail feathers and drooping wings, she will engage in courtship and copulation with

the male—then she will leave him forthwith to visit other males in the vicinity. When she is ready to deposit her first egg, she will go to the next cock who calls to her, leave the egg with him, and take off. When she is ready to ovulate again, she will head straight for the nearest male in the vicinity and offhandedly deposit the egg with him. Thus, each cock ends up with a nestful of eggs laid by different hens, which he is left to care for completely on his own. Tinamou eggs, incidentally, are among the most beautiful of all birds' eggs. They are chestnut brown, burgundy red, purple, green, or blue, depending on the species, with a lustrous, lacquer-like gleam. The cock leaves the nest but once a day, generally for a short time only, to look for food. During these brief absences, some species simply let their eggs lie conspicuously in the open, while others will cover them, more or less carefully, with leaves.
This peculiar breeding behavior is exhibited by most tinamous. But in some species, the cocks will accept and care for only one egg, and in still others, the cocks will apparently hatch the entire brood of a single female, consisting of from four to nine eggs.

The perfect form of polyandry also seems to exist among other birds: it is practiced by the jacanas, a bird family whose outward appearance and way of life resemble the rails, although they actually belong to the sub-order of plovers. All jacanas probably exhibit a similar breeding behavior; consequently, that of the Indian pheasant-tailed jacana, which has been carefully observed, may be considered as typical.
The male pheasant-tailed jacanas build floating nests at breeding time. The female will visit one of the males, deposit four eggs in his nest, over a period of approximately ten days, and then be off, leaving the male to brood devotedly—and unaided—on her eggs. Meanwhile, the promiscuous female will search out the next nest owner, mate with him and, in due course, will bless him with four eggs of his own. This process will often be repeated twice more (two to four males for every female is the general rule). Nor does it end there: the eager female barely waits until her first mate has seen his brood to independence

before depositing another four eggs in his nest. Her other mates will be visited and blessed in like manner, and thus the busy female will frequently produce eight sets of four eggs in the course of each breeding season.

COURAGEOUS GRASS WIDOWERS

Emperor penguins practice a peculiar form of mating. These large birds—they are almost forty inches tall—live in Antarctica along with a number of other species which are especially inured to the cold. While all the other inhabitants of Antarctica—like most bird species—take advantage of the warmest season for breeding and raising their young, the emperor penguin does his courting in the southern autumn when the days grow shorter. He lures his partner with the orange-yellow sides of his head (in cases where these were experimentally dyed dark, it was found he was doomed to remain single).

When the murderous winter storms break loose and most other birds have migrated toward the northern spring, the female emperor penguin lays her egg, which her mate immediately rolls on his foot and surrounds with a warming abdominal fold. The female, having done her duty for the time being, then waddles back to the sea, which is often a great distance from the breeding site. For approximately sixty-three days, from late November until early February, during the perpetual darkness of the polar winter, the male penguins stand huddled close together protecting their eggs, defying snow, storms, and brutal cold. The females do not return from the sea to rejoin the grass widowers until the young are hatched. Finally, the males, who have lost about thirty percent of their weight during the long brooding season, can return to the sea to find food. Counting the courting period, they have fasted by now approximately three months without interruption.

At the age of five months, the young can look after themselves, and the adults have time to moult. As is customary among penguins, they lose all their scalelike feathers almost simul-taneously, and thus they cannot go into the water for six weeks. This forces them once more to fast, using up their fat reserves.

FAITHFUL UNTIL DEATH

Just as geese and large parrots live gregariously in big flocks, at the same time staying together as couples for many years, crane partners also remain faithful to each other for life.

Northern cranes return to their breeding areas in the early part of the year. Sometimes they will arrive while deep snow is still covering the ground. Each couple will soon select a nesting site which is invariably near the one of the previous year. Interestingly enough, this holds true even in cases where they have been disturbed at that site the year before to such an extent that they were unable to carry out the breeding task successfully. This persistence in holding on to a breeding area may be one of the reasons why cranes are in such jeopardy. One third of the fourteen species is already seriously threatened. Only about fifty specimens of the North American whooping crane are still in existence, but thanks to the most rigorous conservation measures, they may yet survive; at least, their number has increased again recently. In 1941, only seventeen of these magnificent white birds were alive.

The African crowned crane is the only species which occasionally builds its nest in a tree; all others breed on level ground. Some will deposit their eggs, usually two in number, on the bare earth, while others will assemble a considerable pile of plant debris with which to build a nest.

Male and female share equally in looking after the nest and tending the eggs. The couple starts brooding immediately after the first egg has been deposited. The second egg is not laid until two days later, and, consequently, the young hatch at a two-day interval. The chicks are already able to walk on the first day, and soon after the second egg is hatched, the couple will leave the nest with their young. The parents will then start molting forthwith. While the majority of birds shed their pinion feathers gradually and thus remain capable of flight at all times, cranes, for the most part, lose all their pinions within two days and must therefore remain on the ground for about six weeks. These feathers are only molted every second year, however, while the remainder of the plumage is changed annually. The wings of the young are fully developed when they are approximately ten weeks old, by which time the parents, too, can fly again.

In autumn, the northern crane species assemble at old-established locations before departing on the long voyage south, flying in magnificent wedge formations, accompanied by loud whooping which serves to keep the flight together. Even within the formation, the whole family will remain intact. In spring, the young cranes will leave their parents, but they will not seek a mate of their own until they are four years old. Under favorable circumstances a crane marriage may last fifty years.

SOLICITOUS FATHER IN FULL DRESS

When a male bird parades a resplendent mating plumage—complete, in some instances, with eccentric long feathers—before a plainly feathered female, it is fairly safe to assume that he will take little or no part in the care of his offspring. But once again it is a rule with certain exceptions, the most conspicuous and famous being, without a doubt, that of the quetzal. The males of these crow-sized birds have a brilliant red abdomen, while most of the rest of their plumage is tinted a magnificent metallic iridescent green. The delicately frayed feathers on their backs are draped laterally over the wings when at rest, and the tail coverts form a marvelous train, which may be more than forty inches long. The quetzal's plumage can compete with those of the most beautiful male pheasants or birds of paradise. Nevertheless, he assists with hatching the eggs and feeding the young, and, also despite his resplendent plumage, he is monogamous, at least for one breeding season. He is, however, a troglodytic breeder; in an open nest, he would far too readily betray his own presence and that of the eggs. During the mating season, the quetzal and his far less conspicuous partner look for a rotting tree trunk. Together, the couple will construct a breeding cavity similar to that of a large woodpecker. But since their short bills are not suitable for chiseling, they will actually gnaw the cavity into the soft wood—which accounts for their designation as trogons, or gnawers. Into this bare cavity the female will generally deposit two light blue eggs, about as large as ping-pong balls, and almost spherical in shape, like owls' eggs. Every morning and every afternoon, the male will regularly relieve his wife at the brooding task for a few hours, paying no particular attention to his magnificent train. Indeed, after a short time, the feathery ornamentation becomes sadly disheveled, and only after his next molting will the quetzal resume his previous splendor. The young, which are initially naked and blind, hatch after an incubation period of eighteen days and for another month they are fed by both parents, at first with insects, and later with small vertebrates and fruit as well.

THE INGENIOUS INCUBATOR

Mound fowl practice one of the oddest, most amazing propagation methods of all birds. Twelve different species of these gallinaceous birds inhabit New Guinea, Australia, and several of the neighboring islands. As is true with many gallinaceous birds—as well as with other ground breeders—mound fowl dispense almost entirely with the family unit. Fortunately, their fledglings require almost no care. Moreover, the brooding is done by one parent, and hence a "conventional" union is superfluous. The young of the mound fowl are early fledglings and extremely independent. From the very first day, in fact, the young mound fowl must begin to provide for their own food and see to their own safety, since neither of the parents will look after them. Among most

mound fowl species, the incubation of the eggs is an ingenious, and at the same time complicated and time-consuming procedure. For these peculiar birds hatch their eggs by incubation, and must select—or construct—an "incubating oven" with just the proper temperature for the process. Most scrub fowl accomplish their task with relative safety and ease. Their main habitat is the Solomon Islands, which are of rather recent geological origin. At breeding time, the scrub fowl will look for a region where the sandy ground is still warm from recent subterranean volcanic activity, and there they will bury their eggs at a depth of up to forty inches. Once they have deposited their eggs in a suitable place, they cover them and leave them to their fate. Their accurate sense for the correct incubation temperature—approximately 93° F—is uncanny. In the areas they select, ground temperatures are so constant that a large proportion of the embryos will develop without further attention on the part of the parents. Especially popular as incubating sites are areas in the vicinity of hot springs, where the proper temperature and humidity are close to the surface of the ground, thus reducing the labor required for excavation. Not all scrub fowl, however, are able to locate the proper soil at the proper time. But they know how to help themselves: they build incubating ovens, mounds with diameters of up to forty feet and heights of more than sixteen feet, making them the largest of all birds' nests by far. In densely forested areas, these mounds will consist of foliage which will supply the proper fermentation heat in the process of decomposition. In open areas, the nest is constructed primarily of sand, and the necessary heat is provided by the sun.

The talegallus build their incubating ovens in the dense jungles of New Guinea. They have no access to volcanic or solar heat and must rely on the heat of fermentation alone, a method they have improved for reasons of safety. The achievements of these birds are almost unbelievable. With his large, powerful toes, the cock will scrape together a pile of foliage measuring nearly ten feet in diameter. When the structure is

finished, the bird will remain in its vicinity. Each day, he will dig through the pile to insure that it will be dampened all the way to the round by the rainwater. Thus an intense fermentation, producing high temperatures, is quickly set up in the humus. At this point, the heat would be far too great for the eggs. Not until later, when the temperature is reduced, will the cock permit the hens to approach the nest and lay their eggs. As soon as they have done so, they are relieved of their duties—but the work of the cock has just begun. He continually scrapes up more foliage, intermingling it with the pile, so that continuing fermentation can maintain a constant temperature. He tests this temperature by digging holes in the humus, thrusting his head deeply into them and opening his bill. (The heat-sensitive spots are located inside the bill, probably on the tongue and on the palate, but the exact location is still unknown.)

Almost all mound fowl live near the equator, which provides them with a more or less constant, high temperature for their endeavors. But one relative of this family, the mallee fowl, inhabits the southwestern region of Australia, which is, indeed, as far distant from the equator as Spain or the State of North Carolina. Not only does this area have considerable seasonal temperature fluctuations, but daily fluctuations as well—within a twenty-four-hour period, temperatures may vary as much as 72° F. Fermentation heat would be welcome indeed, but in this region the mallee fowl are unable to produce it with simple piles of foliage. Few leaves are to be found on the ground in the arid bush, and, in any case, the extremely sparse rainfalls during the summer months, combined with frequent, drying winds, are anything but ideal conditions for decomposition, even if large quantities of foliage are available.

Nonetheless, mallee fowl do rely on fermentation to incubate their eggs, and consequently the mallee cock has to perform an extraordinary amount of work. In autumn, he begins the construction of his incubator. He scratches a pit, about forty inches deep and eighty inches in diameter, into the ground. When he has finished, and

*Throughout the day,
the swallow-tailed gull parents remain
in the immediate vicinity of the nest,
in order to defend the egg or chick
against the frequent attacks
of the nest-robbing frigate bird.
Their large eyes gave rise
to the conjecture that they go
in search of food during the night.
This was actually confirmed recently
and is probably a unique trait among gulls.*

the first scanty rains of winter set in, he must proceed quickly in order to exploit the slight quantities of precipitation. He will collect all the foliage and every dry twig within a radius of about 170 feet and throw it into the pit. Next, in order to retain the moisture of the few rainfalls in the leaves, he fills the hole with sandy earth and above that throws up a sandpile which may be over forty inches high and more than sixteen feet across. While he is working, he must constantly rout other cocks from his area, because he needs all the nesting material and food it will provide for himself.

Finally, in about four months, the incubating oven will be finished and the necessary fermentation heat attained deep down in the pit. The hen can now start laying; but instead of producing her eggs in a twenty-four-hour cycle like most other birds, she lays them at intervals ranging from five days to more than two weeks, producing about thirty eggs in one breeding season. Seven months may pass from the first ovulation until the last chick has hatched, during which time the drudging cock must constantly watch over the nest, an activity requiring almost as much labor as its original construction. He scrapes about his incubator an average of five hours a day, testing the temperature of the breeding chamber with his bill, and adding to or taking from the sand layer covering the eggs. In spring and autumn, the outside temperature will be far below the required incubation temperature, but during a summer day, it may rise more than 18° F above it. To keep the chamber temperature constant, he will pile sand heated by the noonday sun on the eggs during the cooler periods, while on summer mornings he will cover them with surface sand that is still cool from the night. He can also regulate the temperature by means of cooling shafts which he digs into the pile.

The technique of the mound fowl does not appear to be particularly practical; the mallee cock works eleven months out of the year to obtain the same results which other birds are able to achieve within approximately two months. But then, who can fathom the many peculiar breeding habits of birds?

A Permanent Home for Two

Storm petrels, sea birds, related to the albatross, brood in the concealment of protective recesses—either earth caves which they dig themselves, or ready-made rock cavities suitable for their purposes.

It is interesting to note that young storm petrels seek out a breeding cavity of their own when they are only two years old, although they will not be capable of propagation for another year or more. Once in possession, they will, as a rule, turn to the same cavity year after year. Out of season, storm petrels migrate thousands of miles away from their breeding grounds, but at mating time they will return, each couple unfailingly identifying its own home.

Both partners alternately sit on the single egg. Some species relieve each other at rather long intervals of a few days, during which the brooding partner is sometimes supplied with food by its mate. Approach to and departure from the breeding cavity generally takes place only under cover of darkness. After approximately six weeks of incubation, when the naked, blind, and helpless chick emerges from its shell, both parents will tend to its needs, feeding it with an oily, extremely nourishing liquid. In about seven weeks, when the nestling has reached a weight nearly double that of its parents, it will be abandoned, and for almost two weeks it must live on its vast accumulated fat reserves. During this period in which it will loose about half its weight it develops wings which are fully capable of sustaining flight, so that, immediately upon leaving the nesting cavity, the young bird can soar confidently into the air.

Switched Roles

Among most birds, the females do the main work of brooding and raising the young. Many do so even without any assistance from their males. But, as we have already seen, there is a considerable number of species whose males look after eggs and offspring on their own. Almost none of them, however, have switched roles as completely as the phalaropes, a small group of dainty bird species related to the plovers, living in the northern tundras of the Old and New World. In most bird species, the males have—to a greater or lesser extent—a more colorful plumage than the females, or else they are similar in coloration. Usually, the males also are the first to arrive in the breeding area after leaving the winter quarters, and play the more dominant role in courtship when the females arrive a few days later.

In the case of the phalaropes, however, a considerable quantity of male sex hormones in the ovaries of the females has upset all of these general rules. In their breeding plumage, the female phalaropes are far more colorful than the males, and, as a rule, they arrive in the breeding area several days in advance of the males. The mating process also takes on a rather twisted aspect. When the inconspicuous males arrive, each female will select one of them and remain constantly in his vicinity. Woe to the unattached female who attempts to play up to a male that is already taken! She will be angrily attacked and driven off by his mate in an aggressive dominance struggle that is normally the custom among males. Also in the courtship ritual as such, it is evidently the female who courts the male with distinct gestures. She will deposit her four magnificently camouflaged eggs in a depression which she makes in the sandy ground with her breast by twisting her whole body. Then the male, ignominiously deserted by his mate, settles on them with docility and broods. She may then look for another partner, though there is not conclusive evidence on this point.

THE BIRD FAMILY

Most bird species producing late nestlings prefer the greater protection of bushes and trees for their nesting sites. Troglodytic breeders like the blue titmice are among the particularly well protected species.

Whether the first sensation of soft, wriggling life under its body comes as a surprise to the brooding bird, is, of course, a moot question. Certainly, the young often begin to peep inside of the egg as early as the second day prior to hatching, thus establishing vocal contact with their parents. This is answered by the older birds and probably prepares them for their new problems. These calls from within the egg may even arouse the feeding instinct of the parents. Rails, for example, can be observed attempting to feed the peeping eggs.

With the hatching of their offspring, the tasks faced by the parent birds vary greatly. In the extreme case of the mound fowl there are no tasks at all, because the young can look after themselves from the first day on, and among some species can fly immediately. But other bird parents have to take care of their young for at least twelve days, usually for a few weeks, and frequently for several months.
The newly hatched young fall into two categories: early fledglings and late nestlings. Early fledglings are able to use their

Patiently
and carefully,
the eggs
are hatched.
The great task
has just begun;
for now
the tiny, helpless young
must be protected,
and hungry, gaping beaks
be fed.

The cradle of the willow warbler's young rests on the ground, but is excellently concealed by vegetation.

feet from the first day on. To be sure, some can barely stand on them or can walk only very slowly, but all of them, with the exception of the penguins, are at least capable of moving a short distance in case of emergency.

They always have a protective down covering, sometimes of spotted camouflage coloration, and in general do not need to be fed but are able to peck their own food, which the parents often have to point out to them.

Late nestlings always appear helpless by contrast. Most of them are naked and blind when hatched, and seem to consist exclusively of a huge belly, big head and enormous gaping throat. It is interesting to note, however, that the late nestlings of most species, as clumsy as they may be at first, achieve full independence more rapidly than the early fledglings. Penguins, ratites, grebes, geese, ducks, gallinaceous birds, cranes, bustards, rails, plovers, gulls, and terns are early fledglings. Late nestlings include all songbirds, woodpeckers and crows, parrots, pigeons, birds of prey, herons, pelicans and their relatives, albatrosses and other tubenoses. Some characteristics of the appearance and behavior of the two categories cannot be clearly differentiated. The denominator common to the early fledglings is their ability to run or swim relatively greater distances—even if they only make use of it in emergencies.

The little coot chick ventures to the edge of the nest when it is only a few hours old. In case of danger it leaves the nest immediately and hides in the reeds.

One of the parents lures the day-old chick into the water, while the other partner sits on the remaining eggs and warms the newly hatched young.

Only a few days old and in the protective company of their parents, the coot chicks begin their rather extensive excursions about the neighborhood, but still remain close to the reed bank.

Although two months old and almost as big as their parents, the young are still fed with aquatic plants brought to them from the bottom.

On the following pages: Most bird species whose young are late nestlings are monogamous; both mother and father are equally concerned about their offspring. During the first three or four weeks after hatching, one parent or the other always stands guard over the small cattle egrets (left), providing them with warmth or shade as needed. But soon the voracious brood demands the all-out efforts of the parents in hunting insects. Nearly fledged, the apparently insatiable offspring of the great spotted woodpecker (right) waits near the entrance to the nesting cavity for the arrival of its parents with larvae rations.

One of the swallow-tailed gull parents always remains near the chick to protect it from the scorching equatorial sun and from rapacious frigate birds. ▶

The young of rails—which include the coots—and of gulls are early fledglings, which is to say, they are already covered with downy fuzz and can run and swim within a few hours. Their behavior, however, is unusual for early fledglings. Rails generally will not leave the nest until one or two days after hatching, after which time they will keep returning to the platform for some time to rest, to dry out and to be warmed. While the little gulls will turn their backs on their nesting trough after a few short days of life and, as a rule, will not return, they will, nevertheless, remain in the immediate vicinity of the nest until they are able to fly, which may take several weeks. Here they will remain, protected by the low vegetation from which they will emerge only when their parents entice them with food. On the Galapagos Islands, however, the parents of the swallow-tailed gulls remain with their offspring for some time, because their breeding sites are located on bare volcanic rocks without protective plant growth.

In bad weather, young barn swallows often demand food in vain, because their parents can only catch insects in flight—and these are in short supply when it rains. Continuous rain can cause a brood to starve to death unless they are fortunate enough to be housed in an animal barn where flies are available at all times.

Only one stimulus
seems to animate
these blind,
dependent creatures:
hunger.

A throat of unbelievable dimensions, an opening with vividly colored, bulging edges that is but faintly reminiscent of the future shape of a bird's bill: there are the most prominent features of the newly hatched nestling. This gaping bill and a digestive system that functions at almost uncanny speed are of decisive importance; everything else plays a subordinate role for the time being and exists only in rudimentary form. Those not familiar with the young

When the young oxeye tits have barely recovered from the exertions of hatching, they will open their bills wide at the slightest indication that their parents are approaching with food.

The vibration of the song thrush nest, provided with a peculiar loamy plaster finish on the inside, is the newly hatched young's signal to "gape."

of birds would hardly believe that the tiny extremities resembling useless appendages will mature, within two or three weeks, into wings capable of sustaining flight or into functioning legs; or that several thousand feathers will

sprout from bare skin within a brief space of time, transforming this lamentable, ugly, naked thing into a charming, fully developed bird. Only the two pronounced protrusions at the head of the little creature provide a hint that extremely sharp eyes are in the final stages of development under lids which are as yet stuck shut. The wide-open gape of the newly hatched young exerts a powerful stimulus on the parent birds. The throats of different species vary in color; usually they are orange or red, with yellow or white protruding edges. In addition, some species often exhibit characteristic patterns, such as black spots in a particular arrangement, or bright blue or red warts in the corners at the base of the bill. The young of several cave-breeding species have especially garish markings to show the way for the food-providing parent in the dark. Some even have slightly phosphorescent spots near the edge of the bill.

The parents react to these stimuli promptly and with indefatigable diligence. They fly off again and again, collect a beakful of food and return to stuff the portions down the demanding throats. This continues all day long, al-

most without interruption, the most intensive feeding occurring during the early morning hours and from late afternoon until dusk. Among species of small birds, the parents may fly to the nest an average of twenty times an hour to provide the hungry brood with food, and many species feed even more frequently. In the case of the titmice, as many as four hundred and eighty transport flights have been counted in a single day.

imminent danger of being spilled out. At the end of the second day, a nestling may weigh twice as much as it did at the time of hatching, and within two weeks its weight may have increased thirty times. When the young leave the nest—which occurs within about ten days among ground-breeding songbirds and in about thirteen days among species that build open nests in high locations, and, in the case of the better protected troglodytic

Bottom, left:
Green woodpecker chicks, in common with many other young birds, especially among troglodytic breeders, will shift, carousel-fashion, after each feeding, so that a different chick waits at the cave entrance for each food delivery, which generally prevents anyone's being neglected.

It is entirely possible that an occasional songbird nestling will be thrust down the gaping gullets of the young jays, along with insects, larvae and other small animals. But the jays themselves may in turn fall prey to a crow which has to provide for its own hungry brood.

Considering these quantities of food, it is not surprising that within a short period of time the nest hardly provides enough room for the brood, which is in

breeders, after about three weeks —they are often heavier than their parents. This helps them to overcome the first days following their time as nestlings.

Children Portraits

1

2

6

7

8

12   17 ▼

13   18 ▼

3

4

5

9

10

11

14   19 ▼

15   20 ▼

16   21 ▼

THE BLUE-FOOT FAMILY

Although a warm,
protected nest awaits
the birth of the baby bird,

The blue-footed booby is entirely satisfied with a nest consisting of a shallow trough in the bare ground. His frugality enables him to establish a family on tiny islands devoid of vegetation, and he can dispense with the time consuming search for nesting materials.

A rather large, light-brownish bird was flying towards the island from the sea. He was heading straight for me at high speed. And when he made his landing approach, he hardly slowed down at all. Almost with unreduced speed, merely with a quick forward extension of his bright blue webbed feet, he bounced on the hard ground amidst the sparse vegetation in front of my amazed eyes, whirled in a somersault and made a quacking noise. And while I was still considering how I could possibly assist the severely injured booby, he pulled himself up and looked at me defiantly as if to say: "What's the matter? I always land like this." Then he waddled a few yards to a bare, excrement-spattered piece of ground, looked at the two eggs lying in a slight depression, turned them over with tender motions of the tip of his beak, spread one foot over them like a protective mantle, and settled down to brood.

This was my first encounter with a blue-footed booby on the Galapagos Islands, and, as I later learned, blue-footed boobies do not always land in this strange fashion. I watched several hundred boobies landing without

when the little chick
emerges
into the world,
he is thrust
immediately
into the harsh reality
of life.

noting a recurrence of this odd maneuver. Yet I remain convinced that the name of these birds is eminently fitting. Their clumsy motions, their ungainly, though nonetheless expressive courting and welcoming gestures, along with a facial expression that conveys the impression of perpetual bewilderment, really seem boobyish. Possibly for this very reason there are few experiences that can be more pleasant and more impressive than an extended stay at a booby colony. Those days, among the numerous bird families on a tiny island of the Galapagos archipelago, were among the happiest of my life even though swarms of

tiny mosquitos did their best to cloud this happiness, and though the fishermen found me close to dying of thirst when they arrived to replenish my water supply after eight days instead of the three days that previously had been arranged.

Only two of the nine varieties of boobies do not breed on the ground, or do so only in exceptional cases: the red-footed boobies and the Abbott's boobies. These build their rather simple, flat brush platforms in bushes and low trees. Only the blue-footed boobies, however, seem to be satisfied with entirely unadorned earth troughs. Expectant parents among the northern gannets, for instance, collect a considerable pile of seaweed; masked boobies prefer to decorate their nesting site with small stones; in constructing their nests, Peruvian boobies and Cape gannets use a little flotsam along with large quantities of their own excrement, which subsequently are used by man as a valuable plant fertilizer.

Northern gannets, which breed on islands in the North Atlantic, deposit only a single egg, while the tropical species usually lay two or, in rare cases, three. All

blue-footed booby nests that I have observed also contained two eggs, as did those I encountered along the coasts of Peru and Chile. The number of eggs laid by a booby is determined by the peculiar circumstance that all of these birds lack "brooding spots": those bare, warm skin areas, located on the abdomen, which nestle against the eggs during incubation. It is because of this deficiency that boobies will cover their eggs with the web of one foot before settling on them.

The brooding behavior of the blue-footed boobies on the Galapagos Islands is no less peculiar. While the birds along the coast will usually hatch both eggs, those on the Galapagos will sit on the nest only long enough to let the first egg hatch. Evidently, the second invariably is left to die. I have not seen a single blue-footed booby family there with two chicks. Presumably, they have become adjusted to living conditions on these islands which are located more than six hundred miles off the mainland: the dangers are fewer here; the life expectancy, the chance for survival of the nestlings is greater than along the coast. By all

This visit of a larger chick from next door evidently is not a mere neighborly gesture: the bigger bird tears at the beak of the little one, then pushes its own beak down the gullet of the little fellow at the first opportunity, causing its victim to regurgitate the contents of its stomach, then feasting on them.

appearances, the boobies practice instinctive birth control by abandoning their second egg.

It is only the method which is unusual, for birth control as such is no rarity in the animal king-

Boobies use their stomachs to transport fish for their young to the nesting site. To obtain its food, the little one thrusts its whole head into the wide open beak of the adult bird, somehow managing not to injure the gullet of the older bird in the process.

The young booby is blind, naked, black, and ugly when it emerges from the egg. It opens its eyes after a few days and, after about a week, sports a fluffy, snow-white down covering. A substantial protein diet helps it to grow rapidly; two months after hatching, it wears its full adolescent plumage and is as big as its parents and considerably heavier. Now it is abandoned by its mother and father, but will remain near the nesting site for another two weeks or so, begging from strange boobies without success, and now and again securing a meal from younger members of its own species by brute force. Then, when it finally ventures to fly down to the water, it will have to fast for more than two more weeks while swimming about, before being able to take wing and dive for its substance.

dom. Many mammals, especially the predaceous animals, will give birth to larger litters in good years. Some birds will vary the number of their eggs; even more frequently, they will increase the number of broods in a single season if the environmental conditions—climate and food supply—are especially favorable. And while there are other bird species which frequently abandon a portion of their eggs (great crested grebes are an example), they do so for a different reason. The first four or five grebe chicks that are hatched seem to cause their parents so much work that there simply is not enough time left to continue brooding. It is most probable that environmental conditions in no way influence the great crested grebes into abandonment of their remaining eggs. Frigate birds brood in bushes within the immediate vicinity of the blue-footed boobies, who constantly must be on guard against these predators. Eggs and small nestlings are considered delicacies by the frigate birds. One of the parents must remain with the booby chick at all times, at least until it is about a week old, in order to defend it if necessary. And even despite this precaution

one of the black birds will snap up a chick with lightning speed in passing flight before the very eyes of its parents and toss the struggling creature about in midair, playfully catching it again and again before finally devouring it. The adult boobies also are frequently molested by the frigate birds when they fly home from a fishing trip; so much so that they will vomit up the heavy contents of their stomachs to make themselves more maneuverable. (That is precisely what the robbers want. Before the regurgitated fish can even touch the water, these brilliant flyers will snap them up.) The boobies are often injured during these attacks and sometimes even dislocate a wing, drop to the water and starve miserably to death.

Even among boobies there is no easy road to success. Fishing is a difficult task which must be learned the hard way. While young cormorants (which are related to the boobies) continue to beg meals from their waterborne parents until they have become perfect divers and underwater hunters in their own right, young boobies must learn to fend for themselves.

Cormorants look for their prey while swimming on the surface, and then dive down into the water to pursue it. Their plumage does not shed water but becomes soaked; furthermore, the specific weight of cormorants is rather high, and therefore these birds have a considerable draft in the water and need to overcome only a slight buoyancy when diving. Not so the boobies. In common with the ducks, they float on the water like corks. Their fat-im-

rare, and eventually will be abandoned. Though the little one has its full plumage by now, its wings are as yet incapable of sustaining it in flight—the baby fat must be starved off first.

When, after weeks of fasting, the time has come for the young booby to make its first fishing attempt, it will rise into the air, instinctively taking off into the wind. It will then look for prey from a flight altitude of thirteen to seventeen feet and, upon sight-

The family life of the boobies is extremely attractive and vivacious. The movements of these birds, which serve for mutual communication, are especially expressive and are augmented with marked acoustical expression. The "language" of the boobies reminds us in many respects of the communications gestures of the albatrosses discussed earlier.

The feet, resplendent in their brilliant blue coloration, seem to play an important part in the mating game. The courting birds display this peculiar ornamentation as conspicuously as possible as they waddle about histrionically, with tails raised high, bowing to each other from time to time with half unfolded wings.

pregnated and hence air-filled plumage and the numerous air chambers under their skin and inside their bones make it impossible for them to execute surface dives successfully.

When the young booby is eight or ten weeks old, visits during which its parents supply it with a fish ration become increasingly

ing a target, will fold back its wings and plunge into the water. It would certainly be a stroke of luck if it actually were to manage a catch in this way; its speed and momentum are still far too low for effective action. But it will gradually learn to climb higher and to execute the dive with more power. After a few lean weeks of

On the following pages: The booby parents go in search of food individually. Therefore, it is relatively rare that they meet at the nesting site. At such times as they do, they perform a touching greeting ceremony, craning their necks toward each other, uttering quacking sounds which often convey the impression of contentment, and lightly striking their beaks together sideways.

One of the parents will always remain near the nest to guard the youngster, to warm it during the cool hours and to protect it with spread wings against the hot noonday sun until the nestling is about ten days old, fully downed and too big to fall prey to a frigate bird. If it is left alone—even on the first day of its life—it will leave the nesting trough to look for a shady spot under a nearby bush.

Soon the newly hatched chick will undertake rather extensive expeditions throughout the neighborhood, and will beg from every adult bird it meets. Booby parents, however, readily recognize their own offspring and, as a rule, refuse to feed strangers. The piteously begging offspring will often be kept waiting a long time, even by its own parents, who extensively groom their plumage before surrendering their fish.

apprenticeship, it will no longer be inferior to its parents in any way. Now it will let itself plummet like a rock from a hundred or a hundred and thirty feet in the air, break through the surface of the water without slackening its speed, pursue the fish with beating wings and paddling feet at a depth of more than sixty-five feet if necessary, and devour its prey even before resurfacing like a cork. Its particularly strong skull structure as well as the air cushions under its skin help the booby to withstand the shock of striking the water at high speed. Even slippery prey is unable to escape from the serrated edges of its powerful wedgeshaped beak.

CAMOUFLAGE

Among early fledglings, life or death is often a question of camouflage. Chicks of most precocious species are covered with spotted down when they emerge from the egg. Cranes, ducks, and other birds lead their young to hiding places when danger threatens, but members of numerous species will take wing and sound urgent warning calls when an enemy is near. At such times the chicks instinctively flatten themselves against the ground, remaining absolutely motionless until the danger has passed. Usually, they blend so well with their environment that it is almost impossible to make them out.

The characteristic coloration of the chicks of a given species is generally adapted to their habitat. Young rails, for example, are dark—some are almost black—and merge exceptionally well with the dark shadows of swamp vegetation. Desert-dwelling sand grouse chicks are sand-colored. The young of the little terns are marvelously similar in color to the light grayish-brown of the beach pebbles.

Raising a Family

Feeding the Young

The loons reproduce among the lakes and ponds of the tundra regions in the Old and New World. They nearly always look for an island close to the shore on which to build their simple nest of plant debris. While grebe couples will alternately hatch their four to seven eggs, covering the brood with nesting material during their absence, the female loon, which lays only two eggs, warms them by herself and leaves them uncovered while looking for food. The young hatch after about twenty-eight days and leave the nest no later than the second day after emerging from the egg. They are excellent swimmers from the very start and, when only three days old, are capable of evading danger by diving. It will be a long time, however, before they are capable of feeding themselves. Their food, which must be provided by the parents until autumn, includes aquatic insects and the like at first, but later will consist exclusively of small fish. It may well happen during this period that the fish become scarce in the breeding area. In that case, the parents will fly to another body of water in search of food for their hungry young.

Once airborne, the loons are fast, tireless flyers, but each take-off calls for considerable effort. Slapping the water with their wings and kicking simultaneously with both webbed feet, they laboriously attempt to break away from the surface and they require "runway" distances of up to 650 feet. Once airborne, they are very slow to gain altitude, rising only about sixty-five feet in the course of transversing some three-fifths of a mile.

A Curious Method of Transporting Water

Sand grouse, at least the majority of species, are desert dwellers. In their size, appearance, and habits, they are strongly reminiscent of pigeons, but they differ in various developmental characteristics. Their young, for example, are true early fledglings, covered with down and immediately capable of pecking food from the ground. Newly hatched pigeons, on the other hand, are naked and clumsy, and must be fed by a rather interesting method that we shall look into later.

One of the more remarkable characteristics of most sand grouse is their tendency to locate their feeding grounds at a very great distance from the nearest watering site. Since they live primarily on dry seeds, they are unable to satisfy the greater part of their fluid requirements from their food, like most other desert animals. Every day, at the same time (and their punctuality is astonishing), they fly in a group to drink at the water hole. The nesting trough, containing two or three well-camouflaged eggs, may be more than twenty miles from the watering place. As independent as the chicks may be, they are not yet able to cover such a distance. But they cannot do without water either, of course, and therefore the necessary fluids must be provided by their parents.

In former times it was believed that the parents would carry the necessary water to the brooding site in their plumage, but skeptical scientists later rejected such claims as being fantastic. It was assumed instead that the crop served as water container, a theory that can be found in relatively recent ornithological literature. Only a short time ago, however, two investigators verified that the earliest theory was by no means as far-fetched as it seemed; sand grouse, specifically the males, do indeed transport their young's drinking water in their plumage. For this purpose they will wade in shallow water, spreading their abdominal plumage and wetting it by moving their bellies to and from. Having thus thoroughly soaked their feathers, they fly back to their homes. There they stand drawn up to their full height in front of the little ones, who use their bills to strip the water off the feathers. How the male sand grouse manages to store the water in his abdominal plumage in such a way that it is neither lost nor evaporated in flight remains a mystery.

Effective Defensive Measures

When hoopoe couples are ready to mate, they will look for hollow trees, suitable wall recesses, rock crevices, or similar cavities. But evidently they are not satisfied with this protection alone. An additional defensive weapon, as startling as it is effective, is kept ready in the subdued light of the nesting site. The light-colored, almost white eggs, of which the brooding chamber may hold as many as eight, are hatched only by the female, while her mate procures the food—insect larvae, worms, ground insects, and the like—which he picks out of the ground, from cattle manure and other hiding places with his long, slender, curved bill. Even after the brood hatches at the end of sixteen days, the female remains in the nest, and the hardworking male must feed not only his mate, but all of his young as well for another ten days. Then the space probably becomes too crowded and the demand for food too great, and so the female, too, will at last go in search of food. In the absence of both parents, the little ones can be left to themselves with little likelihood of coming to grief. If an intruding stranger should darken the entrance to the cavity, he will be received with a snakelike hissing and would be well advised to retreat promptly if he does not care to make the immediate and most unpleasant acquaintance of the secret weapon of the young hoopoes. If he persists in threatening them, the little ones will turn, raise their rear ends, and a barrage of excrement will be squirted, accurately and under considerable pressure, into the face of the unbidden guest. At the same time, the dark cavity will exude an odor that simply defies description.

This penetrating stench, almost unendurable to any half-way sensitive nose, is produced by the large preening glands of these birds. Only in young hoopoes and in the brooding female has the oily secretion, normally intended for the care of the plumage, such an intense aroma. Not too much is known about the actual effectiveness of this weapon as yet. Cats, however, will retire with every indication of disgust and without having achieved their purpose when confronted with this smell, and it is entirely probable that other would-be nest robbers lose their appetite for young hoopoes after one whiff.

The Daring First Step

Young murres first see the light of the world on rocky plateaus, or frequently, on tiny projections and narrow rock ledges of the ocean cliffs near the North Pole. They emerge from eggs that are laid on bare rock, but which cannot roll off even a slightly inclined surface because of their pear shape. These eggs, incidentally, bear spotted markings which vary considerably with each individual egg, thereby enabling parent birds unerringly to identify their own egg—even in cases where they have been mixed up with eggs from different broods.

About two days before hatching, the chicks call attention to themselves with peeping noises, and the parents reply. As we have seen earlier, young birds of various species make vocal contact with their parents from inside the egg to prepare them for their imminent new tasks.

The murres are early fledglings covered with fuzzy down. They soon begin to make short excursions and visit the countless offspring of their neighbors (murres breed in colonies of hundreds or thousands of birds), but they do not leave the actual breeding site until about three weeks later. Then, however, they do so in a most dramatic fashion: when the parents feel that the time has come for their young to become acquainted with the sea, their primary element, they fly down to the water, from where they will call enticingly to their children. The little fellows are still incapable of flying, although they already have what ornithologists call their "interim plumage." Moreover, they are only about one-fifth the weight of their parents (who weigh approximately 2.2 pounds) and therefore unfit for diving. Nevertheless, with a daring leap they will dash over the edge of the cliff into the sea, which often lies as far as one hundred feet or even more below. Frequently, they will strike projections or crash onto the gravelly beach below. Injuries, however, are rare, since their bodies are still extremely light and their bones very elastic. Thus, despite the suicidal appearance of some of their abortive dives, the young birds usually shake off their shock and soon will be ready to practice diving and fishing.

The young of some other birds venture similarly audacious leaps into life; newly hatched ducks and whistling ducks, for example, which occasionally are incubated in the hollows of trees at very great heights. When these one- or two-day-old youngsters strike the ground, they bounce like small rubber balls.

Economical Food Transport

Many small songbirds are capable of carrying extremely large numbers of caterpillars, spiders, insects, worms, and the like in their bills; a capability without which they would soon exhaust themselves physically because, in every case, the voracious brood requires an enormous volume of food. Even so, in many species the parents must make more than two hundred daily flights to the nest with crammed bills to satisfy the appetite of their hungry chicks.

But how can a portion of slippery wriggling fish be transported? The simplest method is probably to swallow the prey during the search for food, carry it to the nest inside the stomach, and then regurgitate it. This technique is practiced by many fishing species, including herons, pelicans, boobies, and cormorants.

The fish fanciers among the birds of prey seize their prey with their taloned toes. This method permits them to carry only a single fish at a time to their aeries, but that is generally one of considerable size, and the offspring will take some time to digest it.

Terns carry food to their children in their beaks, again only one fish at a time. This prey, too, is often relatively large, and it is by no means rare to see an infant tern with the tail end of a fish still sticking out of its beak, while the front portion of the prey is being digested in the stomach. Murres, on the other hand, take relatively small fish to their nestlings, one at a time, and so are kept busy flying back and forth until the young are big enough to leave the breeding site.

Several relatives of the murres—such as the razor-billed auks and the puffins, which also are members of the auk family—have developed a particularly odd method of transporting their food. They fill their bills crosswise with as many as a dozen fish at a time and return to the breeding site with their prey hanging down on both sides of their bills. Anyone who has seen a bird with his bill loaded in this fashion will find it difficult to visualize how it managed so many successive catches without almost as many losses. Actually, the bird solves this problem in the simplest possible manner: it clamps the fish against the palate (in the upper part of the bill) with its tongue, leaving the lower portion of the bill free to scoop up the next catch.

MAMMALIAN AMBITIONS

Pigeons are unique among birds: the first food fed to their young, of which there are almost invariably two to a family, is milk.
While the adult pigeon sits on its eggs, a special hormone, prolactin, which is responsible for milk secretion in pregnant mammals, is activated. Its effect is to cause the mucous membrane to swell to great size and begin to secrete a fatty substance which collects in the crop as a curdlike, somewhat granular mash. This "pigeon milk" actually is quite similar in composition to the milk of mammals—in particular, to that of rabbits—but, unlike mammal's milk, it is secreted by both sexes.

The young pigeons, which emerge from the eggs after an incubation period of about eigtheen days, are still clumsy, like newly hatched songbirds, and almost completely naked and blind. To obtain their food, they have to suck while pushing their bills and half of their heads down the open throat of the parent bird. This, incidentally, is another unusual characteristic of the pigeon. Adult pigeons are among the very few birds that drink by sucking water into their bills. Most birds have to scoop up water with their bills, which must then be raised to let the liquid run down their throats. After a few days of this exclusive milk diet, the rapidly growing young pigeons are fed increasingly with seeds, until their unique source of nourishment dries up altogether about ten days after they have hatched.

PROTRACTED CHILDHOOD

We have already encountered the strange reproductive habits of the emperor penguins, who brood during the Antarctic winter. Their young take about five months to grow to the size of their parents and to begin fishing on their own—a very long childhood for birds, most species of which grow up in much less time. As slow as the emperor penguin may be to mature, its nearest relative, the king penguin, takes far more time. The brooding habits of the king penguin are also different in many respects from those of the emperor penguin.
The king penguins live in Subantarctica. They brood in summer; i.e., not in snow and ice, although in a very harsh climate against which they must protect their eggs. For this purpose they have a skin fold, or brooding pouch, above their feet, similar to that of the other large penguin species. The king penguin lays a single egg, which both parents brood alternately. At the time of brooding relief they will stand toe-to-toe and carefully roll the egg from the feet of one to those of the other, so that it does not come in contact with the cold ground. After an incubation

period of nearly two months, the young hatches and immediately is stuffed with food by its parents. It grows rapidly and develops a thick layer of fat for protection against a long winter that will come while the youngster is still far from ready to swim and feed itself. When the storms set in, the young huddle closely together in flocks, surrounded by adult birds who attempt to protect them from the wind. Once autumn has passed, food becomes scarce and the young receive a fish ration only once every two weeks. They will have lost no less than half their weight by the time spring once again affords better hunting—a time when their parents must still look forward to several months of caring for offspring that still do not have their waterproof plumage and will not go fishing on their own until they are about a year old.
Other birds that require similarly long rearing periods, such as the albatrosses, usually breed only every second year. The king penguin, however, manages to breed twice every three years; one year in the early summer and at a later season in the second year, after a few weeks of recuperation from the completed task of rearing the offspring. Finally, in the third year, it will take a well-deserved rest.

PROSPERITY BREEDS NEGLECT

Many birds are capable of adjusting to changing environmental conditions in a very short time. This capability is particularly apparent in the various beneficiaries of culture, those species which have learned to take advantage of man-made modifications of nature. These modifications will enable them at times and under certain conditions to increase their numbers considerably and to extend their habitat.
In this way, the arctic fulmar managed to move the southern boundary of its breeding area from Iceland to the south coast of England within a period of only a few years. It advanced about

twelve miles each year, probably encouraged by the waste products of the fishing industry.

The spread of the cattle egret is even more impressive. In 1963, I saw a group of cattle egrets near some cows owned by Seminole Indians in the Everglades of Florida. I could hardly believe my eyes, since, as far as I knew, there were no cattle egrets in North America, nor were they listed in any handbook on North American birds. I looked into this matter and discovered an almost unbelievable story.

Within the past few decades, this bird was able to multiply vigorously and spread widely throughout Africa, its original habitat. The growing cattle population and increasing agricultural activity improved environmental conditions for the bird. Just as it had followed game animals in the past, feeding on the insects those animals attracted, it later followed domestic cattle, catching flies while walking between the legs and over the backs of cows. Soon it also learned to follow plowing farmers, as crows, gulls, and other birds do in our regions. About 1930, the African continent, huge as it may be, appears to have become too small for the cattle egrets. It was probably a group of young birds—much more prone to migrate than their elders—who crossed the Atlantic, landed in South America, and settled there. They evidently liked their new home, for they conquered the entire northern part of the continent, even adjusting to the murderous climate of the High Andes (where I have seen flocks of them following llamas at altitudes of more than 13,000 feet above sea level) and spreading northward throughout Central America. The first of these birds seems to have shown up in North America around 1953. Today, there are probably myriads of cattle egrets living in the New World. (In the meantime, the cattle egret had emigrated to yet another continent, Australia, where the first of them were observed in 1948.)

Such an enterprising creature naturally is considered an excellent subject for study by the behavioral scientist. Otto Koenig caught several young cattle egrets in North Africa, took them

to the research station at Wilhelminenberg in Vienna, and began his researches by letting them fly at liberty there. In autumn they were recaptured and placed in a heated room, and in the following spring they were moved to a large, newly constructed outdoor aviary. But a curious thing occurred in the interim: the first year-old egrets had already grown the beautiful cinnamon-colored ornamental feathers of their breeding plumage (heron species normally do not grow their breeding plumage until their second year and until then are incapable of reproduction). A few days later, one couple began to build a nest, and after a week, the first pale blue egg was lying in it, brooded alternately by each partner. Additional eggs followed at two-day intervals, and soon the world's first cattle egrets to hatch in captivity actually emerged.

No matter how spacious an aviary may be, its environmental conditions can never be the same as those experienced by birds at liberty. The time-consuming chase and the extensive search for nesting materials are dispensed with, because an abundance of everything needed for existence is available in the immediate vicinity. In the case in point, an unnatural prosperity soon began to have a very marked effect on the behavior of the egrets. In nature, adolescent egrets are forced to look for food on their own within a steadily increasing distance from their aerie during the frequent extended absences of their hunting parents. In this way, they learn to hunt and gradually become independent. Furthermore, the migratory urge of the young birds causes them to resettle in other areas, thus instinctively avoiding inbreeding and all its disadvantages. At Wilhelminenberg, the parents were always present, and food could be begged from them at any time. The young had no opportunity to become independent, and they remained with their parents. Nor did the family break up during the next breeding season. As a consequence, what happened next would never have occured in nature: every male within the family mated at will with every female, all the eggs were deposited in the common aerie, and all the birds tried to brood. There were constant brooding reliefs; frequently

three egrets would sit on top of one another on the nest. The previous year's young were still begging from their parents, obtaining food from them, and giving it to their own children. The courtship and greeting ceremonies, which had been so ritualized before, gradually disappeared. There no longer was any need to become acquainted formally, since father, brother, or son was the marriage partner. Ceremonial greetings became superfluous and were executed negligently or not at all.

In the following spring, this "neglect through prosperity" progressed even further. Various families of the colony began to be absorbed into larger families in which every male mated indiscriminately with every female. Not much remained of the otherwise rather strict monogamous existence of the egrets. Indiscriminate polyandry and polygamy became the rule. But the extent to which the characteristic behavior of the cattle egrets had been disturbed by this modus vivendi became most manifest in the constantly decreasing birth rate. An ever greater number of eggs was neglected because of the brisk activity, and those young that actually hatched were overfed or suppressed.

There is an undeniable parallel here to human development in modern civilizations, and a particularly instructive example for the comparative behavioral researcher.

Otto Koenig believes that it is entirely possible that in time his egrets will develop new adaptation systems which will correspond to the conditions in their changed environment. It will be interesting to see how the cattle egrets of Wilhelminenberg will solve their problems, and whether, unlike man, they will be able to cope with an unwanted degree of prosperity.

THE HOATZIN

Archaeopteryx, the primeval bird, had small, movable, taloned fingers at the joints of its wings. These presumably were used in climbing among the branches of trees. Rudimentary wing fingers still exist among various birds. They usually take the form of spurlike appendages. The South American hoatzin is probably the only species, where these fingers still play a vital role.

The naked, ugly youngsters usually are hatched in a bush a few feet above a calm body of water. When the chicks are only a few days old and still entirely without plumage, they will let themselves fall into the water at approach of danger and swim or dive to evade their pursuers. As soon as the coast is clear, they clamber back onto the shore, where, lacking camouflage, they would be easy prey if they were unable to grasp branches with their feet and the two sharply pointed fingers on their wings and thus to climb back into their nest "on all fours," as their ancestors probably did before them. As the hoatzin grows older and is better able to elude predators, the "archaic" fingers atrophy.

A BIRD COLONY IN THE PACIFIC

Slightly raised platforms built from their own excrement make up the nests of the guano birds. A straggler, a nearly fledged Guanay cormorant is waiting for its parents—while the other young birds of the colony, by the hundreds of thousands, are already off looking for food on their own. ▶

More birds brood along the Peruvian coast—because of the Peruvian or Humboldt Current—than along any other shoreline on earth.

This cold oceanic current originates in the South Pacific, continues northward along the west coast of South America, is diverted towards the west near the equator and dwindles out in the North Pacific. In the ocean off the coast of Peru, the current is nearly 125 miles wide, has a velocity of about six miles per hour and a temperature of approximately 59° F compared to the 77° F of the surrounding waters.

The cold, strongly saline Peruvian Current carries vast quantities of plankton which furnishes the basic diet of huge masses of fish. These fish, primarily anchovies, in turn constitute the food of enormous numbers of birds. On one small group of islands alone, the Guanay Islands of southern Peru, an approximate twelve million birds will gather during a good breeding season, crowding into a minimum of space, and will catch nearly 4,800 tons of fish daily, an amount equal to about 480 freight cars full!

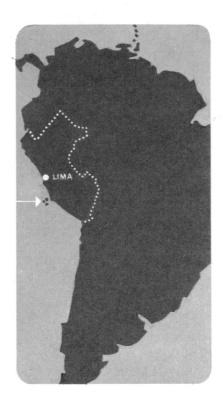

In spite of this great crowding, and possibly because of it, these animals are by far the most valuable wild birds in the world in terms of their usefulness to mankind. The rising currents in this area carry phosphorus compounds from the ocean bottom to the surface, and through the natural food chain, from the plankton to the an-

chovies to the digestive systems of the birds, these phosphates make their way to the breeding and resting sites of the birds in the form of excrement. This excrement, known as "guano," is systematically collected and exported throughout the world as an extremely valuable plant fertilizer, thus providing the Peruvian nation with one of its most important sources of revenue.

The Nino, a warm current from the region of Panama, periodically diverts or superimposes itself on the Peruvian Current at the end of each year. About every seventh year, this north-south current is especially strong, extending all the way down to Chile. The plankton cannot withstand this rise in temperature and are mostly destroyed; the fish population in turn is drastically reduced; and the birds consequently suffer a devastating mortality rate. The beaches are covered with dead and dying birds by the hundreds of thousands. As much as three quarters of the entire bird population may be stricken. And yet, the survivors manage to close these huge gaps within two to three years by increased breeding activity.

A thousand years ago, the Incas already realized the value of the bird excrement deposited on the Guanay Islands. The rulers of the "Golden Empire" caused the guano to be collected there and had their fields fertilized with it. The death penalty was inflicted on anyone who killed guano birds or visited the islands during the breeding period and thereby jeopardized reproduction.

When North Americans, Europeans, and Asians discovered some one hundred years ago what treasures were lying idle on these islands, they began to exploit the deposits—which they encountered in layers up to 130 feet thick—without regard for the breeding periods. In so doing, millions of tons of guano—along with the birds' eggs—were dug away within less than forty years, and were shipped throughout the world.

Only a fraction of the original bird population was still brooding on the islands when the Peruvian government finally became aware of the situation and introduced protective measures. But the guano birds have reproduced with amazing speed, covering the red sandstone cliffs once again with more and more of their chalky white excrement, and today these valuable animals leave nearly a four-inch thick layer of guano, ready for use, on the densely populated plateaus, producing millions in revenue for the state.

Penguins love the cold. The icy waters of Antarctica are their original habitat. The paunchy fellows easily defy the murderous storms of the polar winters. In return, they profit from the vast abundance of fish in the cold seas. Almost all penguins have remained loyal to their native region. Only a few of the eighteen species, seized by the pioneering spirit, have followed the cold ocean currents to the north and settled in new breeding grounds

Thousands of years ago, members of the black-footed penguin species followed the northward urge and migrated with the Humboldt Current. Some settled along the Peruvian coast and, in adapting to their new environment, they evolved into a new species named after their adopted habitat: Peruvian penguins.

Still other black-footed penguins, confining themselves to the current, continued to swim north-

outside the Antarctic zone. All of them have had to pay for this resettlement with a decrease of physical size: those penguins which have chosen the subtropics for their new habitat have evolved into new, smaller species, ranging in height from 18 ½ to 25 ½ inches, while most of the Antarctic penguins are from 27 ½ to 45 inches tall.

ward until they eventually reached the Galapagos Islands, some six hundred miles off the coast of Ecuador. These Galapagos penguins, as they are called, are the only members of their order which breed close to the equator. Occasionally, on their out-of-season migrations, they actually cross it—thus being the only penguins which live, even briefly,

The Peruvian penguins, which formerly inhabited the islands along the Peruvian coast in large numbers and used to play an important part as producers of guano, have suffered from the effects of the perennial guano harvest. While other bird species, notably the guanay cormorant, have once again attained their original huge numbers here, the penguins are encountered now only in small groups.

in the northern hemisphere. Both of these species seek out rock niches and caves as breeding sites. While Peruvian penguins have no fixed breeding season and can propagate in every month of the year, the Galapagos penguins will always mate during the coolest season, i.e. about May or June. Both species generally lay two eggs, which both parents brood alternately. The young do not go into the water until they are about ten weeks old.

The Galapagos penguins, which are only about twenty-one inches tall, are the only species that live near the equator. They breed only during the coolest season of the year, and only on the two southwestern islands of Fernandina and Isabela which are touched by the cold Humboldt Current. Almost their entire day is spent in the water; only at dusk will they climb up onto the volcanic rocks.—Below: Peruvian penguins.

THEY TOOK TO THE ROCK PLATEAUS

Conditions in the region of the Peruvian Current, where there is an abundance of fish, are similar to those in some polar waters. The available and preferred breeding sites—uninhabited islands—are not very numerous, and often

very small. If, therefore, the greatest possible number of birds are to avail themselves of the vast food supply, they have no choice but to crowd together at the nesting sites.

Few species of birds nest in as close proximity to each other as the Guanay cormorant. An average of three excrement-built nests will be jammed into a ten-foot-square area in these colonies —rather cramped living quarters when one considers that these birds are, after all, nearly as large as ducks. Each couple has about two offspring, so that by the end of the breeding season twelve good-sized birds must manage to rest and sleep within ten square feet. Since every couple will occupy and defend its nesting site against intruders despite their gregariousness, incidents occur continuously and spread through large parts of the colony like wildfire. When an approaching parent bird does not land precisely on its own nest, it is subjected to a true running of the gauntlet. Beaks will strike at it from all sides, often felling it, and it must try to rise again as quickly

Pages 178/179:
An ample six and a half feet of wingspread and a relatively light body weight (due to large air sacs under the skin and a lightweight hollow-boned skeleton) make the brown pelican an amazingly fine soarer. These large birds usually brood in self-built brush nests in bushes and treetops. In order to profit from the abundance of fish along the Peruvian coast, however, they must be content with the bare ground of the rocky islands, which are almost entirely devoid of vegetation, and must restrict their nest construction to the assemblage of a little flotsam and a few shreds of seaweed. Several hundred thousand brown pelicans are hatched and raised here almost every year on the excrement-covered ground, in spite of the less than ideal breeding conditions.

Pages 180/181:
The outer appearance of the Peruvian booby differs considerably from that of the large brown pelican, although they are closely related and both belong to the order of totipalmate swimmers. The enormously expansible skin pouch of the lower bill, which is valuable for use as a fishing net, is only rudimentary among boobies. They have compensated for this lack by greatly improving their "thrust diving technique." While brown pelicans (alone among the pelican species) usually fly just a few feet above the water in search of food and barely get below its surface in a short diving flight, the boobies plummet like projectiles from altitudes of 100 to 130 feet and will pursue fish to a depth of some 65 feet if necessary. Like most totipalmate swimmers, they are gregarious and brood in huge, mixed colonies.

as possible to avoid serious injury. It is absolutely uncanny how these birds can identify precisely their own nests in this unholy confusion. By all appearances, however, they are guided by certain individual sounds which only the birds themselves can differentiate, and which assist the partners in communicating with each other.

Pages 182/183:
Guanay cormorant form a gigantic living carpet. Nearly 250,000 birds on this island are crowded together so closely that it is impossible for them to take wing simultaneously. Should any kind of danger cause them to panic, it is quite probable that they would trample each other to death. Once in motion, their feet make a noise like the distant roll of thunder. When they move to their fishing grounds in the morning, the birds on the steep windward shore of the island are the first to take off. This will slowly set the entire carpet in motion as it shifts in the direction of the take-off site, where more birds continually take wing to join the seemingly endless stream which vanishes over the horizon on its search for fish. Three to four hours will elapse from the first takeoff until the last birds are finally able to leave the nesting site.

The greater
the efforts we make
to fathom
the secrets of the birds,
the more our
knowledge expands,
the clearer
becomes
our realization
that the secrets
of the birds
will remain
mysteries of nature as long
as she abounds
with the miracles
of life.

The circle is closing. The young bird has molted. A year has passed, or maybe two or three, depending upon his species. He stands now in his mating plumage, plain or fancy, mute or croaking or bursting forth in wondrous song. He can no longer be distinguished from his parents in any way; he is a fully mature creature.

In his youth he followed his unbridled urge to wander, migrating hither and yon. Finally, he has searched out a home ground of his own, occupying it and vehemently defending it whenever necessary.

His great hour has come. Savagely and full of fire, or tenderly, almost timidly, he seeks to conquer a partner for himself. Having barely shed his fledgling plumage, he is seized by irresistible passion as he pursues his chosen mate. With growing excitement the newly paired birds strive to unite with each other, to attune the ecstatic rhythm of their motions until they achieve that harmony without which there is no procreation.

And then: perfect consummation. The union is complete. The circle is closed.

Once more we are confronted with that great mystery, the origin of life—the beginning of all things. We look at the egg in amazement, and somewhere within us, it must inspire again the great realization: there is no end. The individual is expendable, mortal. But the species will continue to exist, if radical changes in the environment do not stop the natural processes of reproduction. The bird lives on—in the next generation and in all those which are yet to come. It will continue to court, to mate with passionate gestures, to construct a nest with more or less ingenuity, and to turn its eggs cautiously and regularly, hatching them with a patience that is phenomenal in such a lively creature. It will sacrifice itself for its offspring, devotedly looking after the survival of its family, as long as its small bird's heart continues to beat. It will pass on life. It will continue to use its eyes to see and all its senses to receive stimuli from the outside world. And within those mysterious glands will be at work which designate its tasks, arouse its passions, and give it the commands to mate, to build, to brood, to feed.

INDEX OF THE BIRDS IN THIS BOOK

This is an index of all the birds illustrated,
described, or mentioned in this book.
The first column lists the English name;
the second, the scientific name.
Numerals refer to the pages on which each
bird is mentioned
(page numbers in parantheses indicate an
illustration of the bird, its nest, or its egg).
The fourth column shows the distribution of
the various species,
including breeding areas.

Abbreviations:

NA = North America
SA = South America
EU = Europe
AF = Africa
AS = Asia
AU = Australia
AN = Antarctica

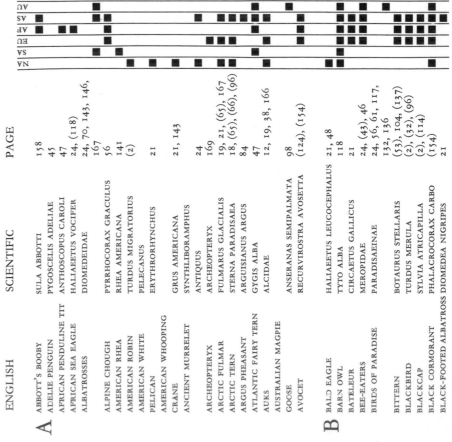

TABLE OF CONTENTS

It is impossible to mention here all those who have contributed to the making of this book, but I am grateful to all of them. There is the anonymous lumberjack in the Peruvian jungle who let me borrow fuel for the outboard motor of my boat, the Scottish fisherman who rowed out to an island despite heavy seas to bring me bread and milk and fish; there are the Kikuyus of eastern Africa who helped to pull my Landrover out of the Mara River, and many others who helped in many ways.

My special thanks go to my wife Monique for her enthusiasm, her collaboration, and her ideas as well as for her criticism.

I also want to thank my indefatigable traveling companions, Max Baumann, Jakob Brauchli, Werner Pfunder, Herbert Gasser, and John Cranham.

And I want to express my appreciation to the following persons and institutions for their contributions toward the realization of this book: Peter Lüps of the Museum of Natural History in Bern; the staffs of the Museums of Natural History in Vienna and Lima, of the Charles Darwin Station on St. Cruz, Galapagos, and of the Institute for Comparative Behavioral Research in Vienna; Roland Wiederkehr and Dr. Fritz Vollmar of the World Wildlife Fund; Dr. Ivan Batthyany of Wildlife Photography Limited; the Ministerio Forestal of Peru; the National Park Agencies of Kenya, Tanzania and Uganda; the National Park Service of Washington; Jack Couffer of Grey Owl Productions; Libby and Jerry McGahan of the National Geographic Society; Verena Hauser, Josef Widmer, Josef VonWil, and Dr. H. Hübscher of the Swiss Society for Natural Sciences.

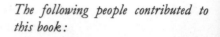

The following people contributed to this book:

Dr. Gerhard Thielcke, Max-Planck-Institute, Radolfzell, expert consultant on the subject of bird calls; Franz Coray, designer of graphic displays, maps and drawings on pages 1–10|11, 12–23, 36|37, 38|39, 42|43, 52|53, 54|55, 68|69, 103, 107, 129, 130|131, 134, 140, 170; Rudolf Küenzi, designer of the illustrations on pages 61, 141, 142|143, 144;
Rolf Baumann, designer of the three color illustrations on pages 42|43; and the following photographers: Lisbeth Bührer: pages 49, 138|139; Karel Hajek: pages 50|51; Paul Niederhauser: pages 30, 32|33, 114|115; James Perret: pages 2|3, 4|5, 6|7 (birds' eggs), 57, 58|59, 60; Walter Tilgner: pages 93, 147|148.

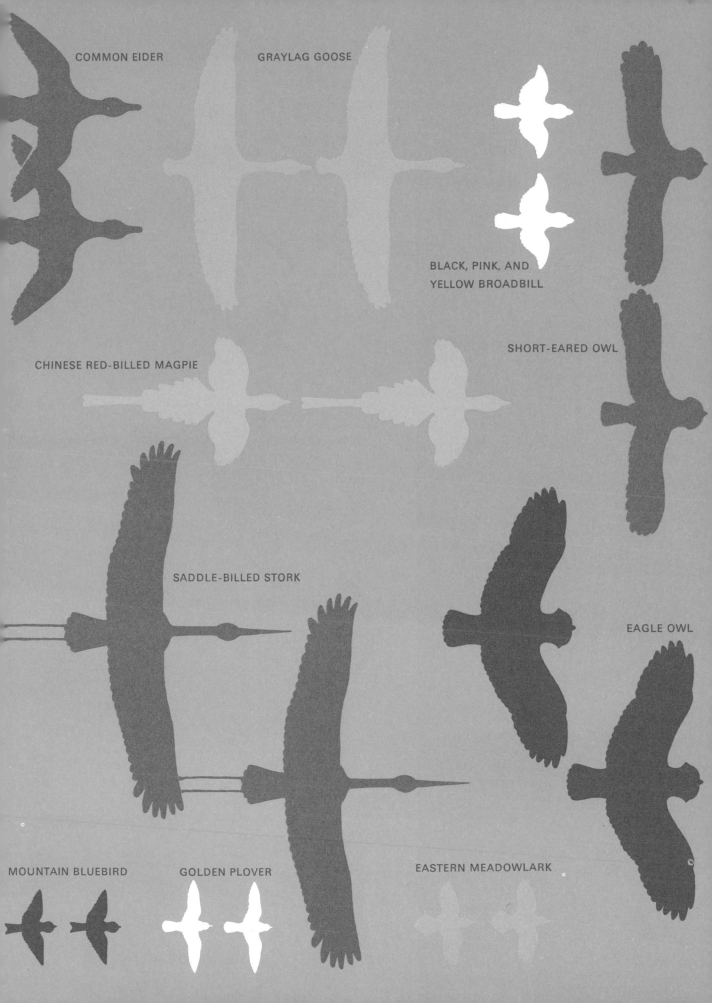

COMMON EIDER

GRAYLAG GOOSE

BLACK, PINK, AND
YELLOW BROADBILL

SHORT-EARED OWL

CHINESE RED-BILLED MAGPIE

SADDLE-BILLED STORK

EAGLE OWL

MOUNTAIN BLUEBIRD

GOLDEN PLOVER

EASTERN MEADOWLARK

AVOCET

BLUE-FOOTED BOOBY

AMERICAN WHITE PELICAN

PIN-TAILED SANDGROUSE

SWORD-BIL

LONG-TAILED WIDOW BIRD

GREATER BIRD OF PARADISE

RUFOUS WARBLER

BLACK-CROWNED NIGHT HERON

BAR-TAILED GODWIT